TRAVELS IN KASHMIR

TRAVELS IN KASHMIR

A Popular History of Its People, Places and Crafts

BRIGID KEENAN

OXFORD NEW YORK DELHI
OXFORD UNIVERSITY PRESS
1990

For the people of Kashmir,
and for Alan, Hester,
and Claudia

Oxford University Press, Walton Street, Oxford OX2 6DP
Oxford New York Toronto
Delhi Bombay Calcutta Madras Karachi
Petaling Jaya Singapore Hong Kong Tokyo
Nairobi Dar es Salaam Cape Town
Melbourne Auckland
and associated companies in
Berlin Ibadan

First published 1989
First issued as an Oxford University Press paperback 1990

British Library Cataloguing in Publication Data
Keenan, Brigid
Travels in Kashmir : a popular history of its people, places and crafts.
1. Kashmir, history
I. Title
954'.6
ISBN 0–19–282791–X

Library of Congress Cataloging in Publication Data
Keenan, Brigid, 1939–
Travels in Kashmir : a popular history of its people, places, and crafts / Brigid Keenan
p. cm.
Originally published in 1989
Includes bibliographical references and index
1. Jammu and Kashmir (India)—Description and travel. I. Title.
954'.6—dc20 [DS485.K24K44 1990]
90–40461
ISBN 0–19–282791–X

Printed in Great Britain by
The Guernsey Press Co. Ltd.
Guernsey, Channel Islands

Contents

Plates and Maps

Acknowledgements

I am grateful to all the people who helped me with this book. In particular I would like to thank Simon Digby, Simon Bazeley, Robert Skelton, Afzal Abdulla, Shameem Varadarajan, Helen Naylor, Veronica Marsh, the Millais family, Anne Gittins, Margaret and Michael Voggenauer, Sandra Boler and Raghubir Singh. I thank the long-suffering librarians at both the London Library and the India Office Library and Records, especially Pauline Kattenhorn. Most of all I thank my husband Alan and my daughters Hester and Claudia—travellers all—for their encouragement.

Credits
Maps by Richard Barraud.
Title-page and fly-leaf decoration by Veronica Marsh.

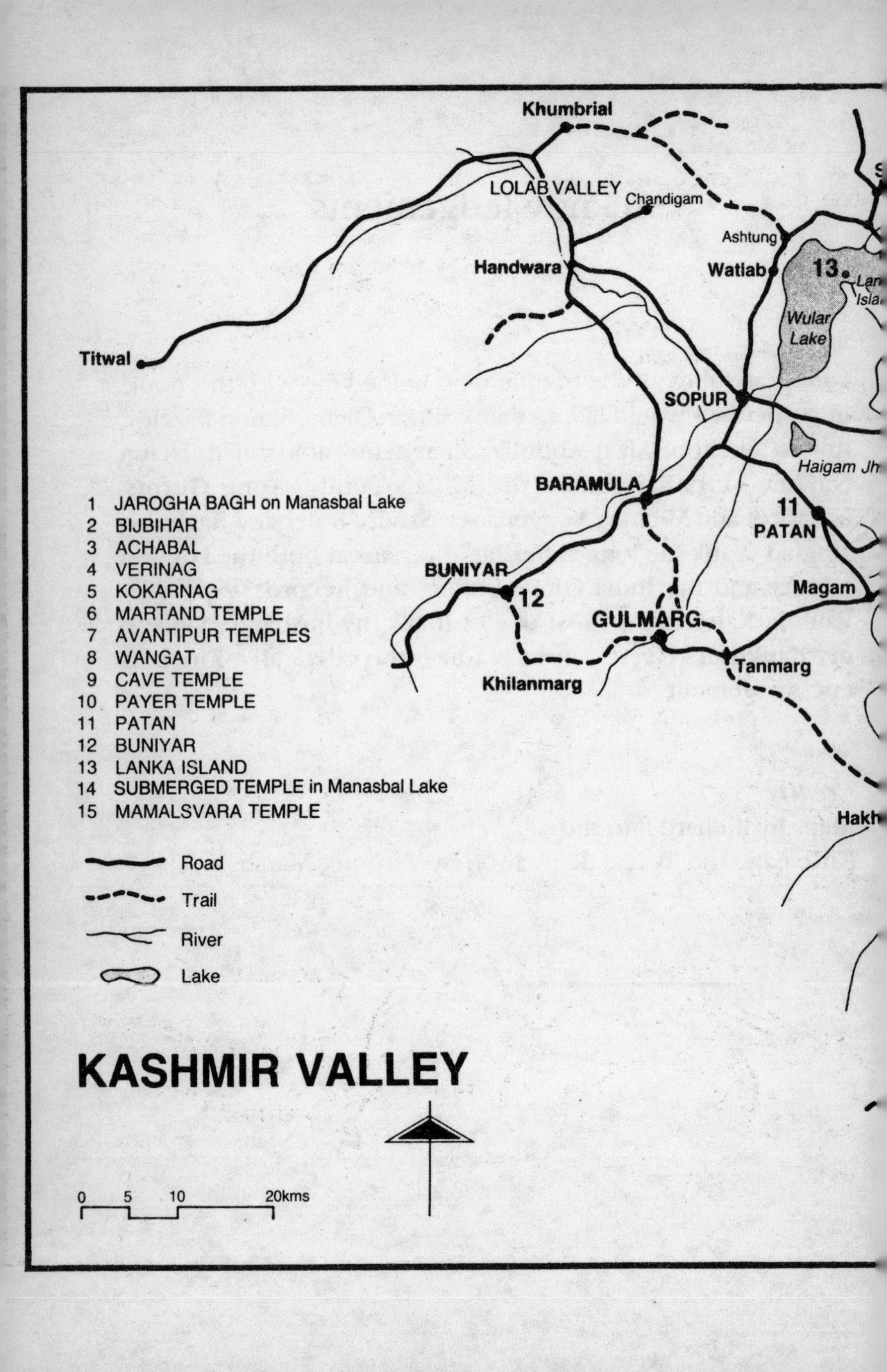

Khumbrial
LOLAB VALLEY
Chandigam
Handwara
Ashtung
Watlab
13.
Wular Lake
Titwal
SOPUR
Haigam
BARAMULA
11
PATAN
BUNIYAR
12
Magam
GULMARG
Tanmarg
Khilanmarg
1 JAROGHA BAGH on Manasbal Lake
2 BIJBIHAR
3 ACHABAL
4 VERINAG
5 KOKARNAG
6 MARTAND TEMPLE
7 AVANTIPUR TEMPLES
8 WANGAT
9 CAVE TEMPLE
10 PAYER TEMPLE
11 PATAN
12 BUNIYAR
13 LANKA ISLAND
14 SUBMERGED TEMPLE in Manasbal Lake
15 MAMALSVARA TEMPLE
Road
Trail
River
Lake
KASHMIR VALLEY
0 5 10 20kms

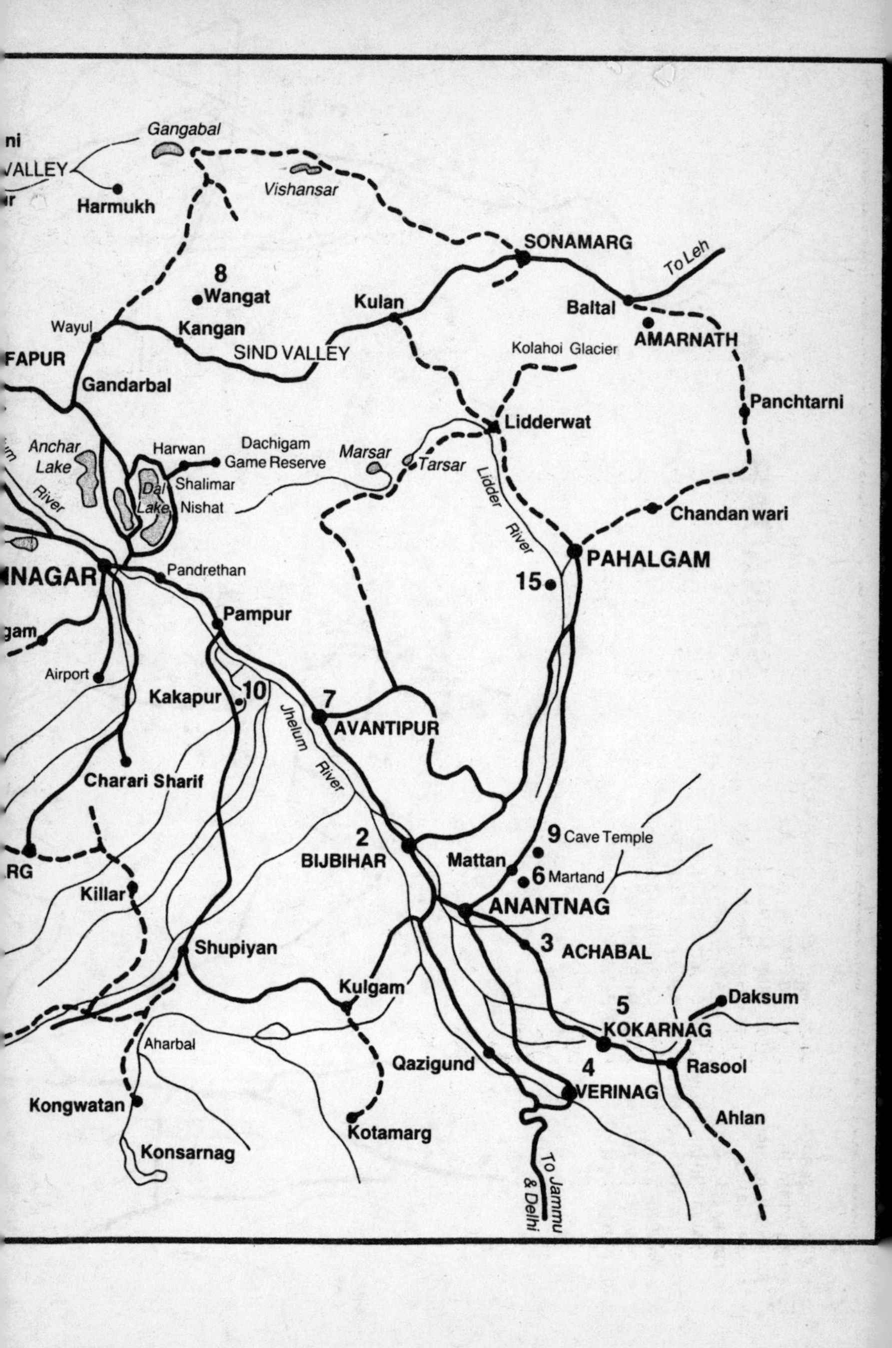

ni
VALLEY
Gangabal
Harmukh
Vishansar
SONAMARG
To Leh
8
Wangat
Kulan
Baltal
AMARNATH
Wayul
Kangan
SIND VALLEY
Kolahoi Glacier
FAPUR
Gandarbal
Panchtarni
Lidderwat
Anchar Lake
Harwan
Dachigam Game Reserve
Marsar
Tarsar
Dal Lake
Shalimar
Nishat
Lidder River
River
Chandan wari
PAHALGAM
NAGAR
Pandrethan
15
Pampur
gam
Airport
Kakapur
10
7
Jhelum River
AVANTIPUR
Charari Sharif
9 Cave Temple
2
BIJBIHAR
Mattan
RG
6 Martand
Killar
ANANTNAG
3 ACHABAL
Shupiyan
Kulgam
5
Daksum
KOKARNAG
Aharbal
Qazigund
4
Rasool
VERINAG
Kongwatan
Ahlan
Kotamarg
Konsarnag
To Jammu & Delhi

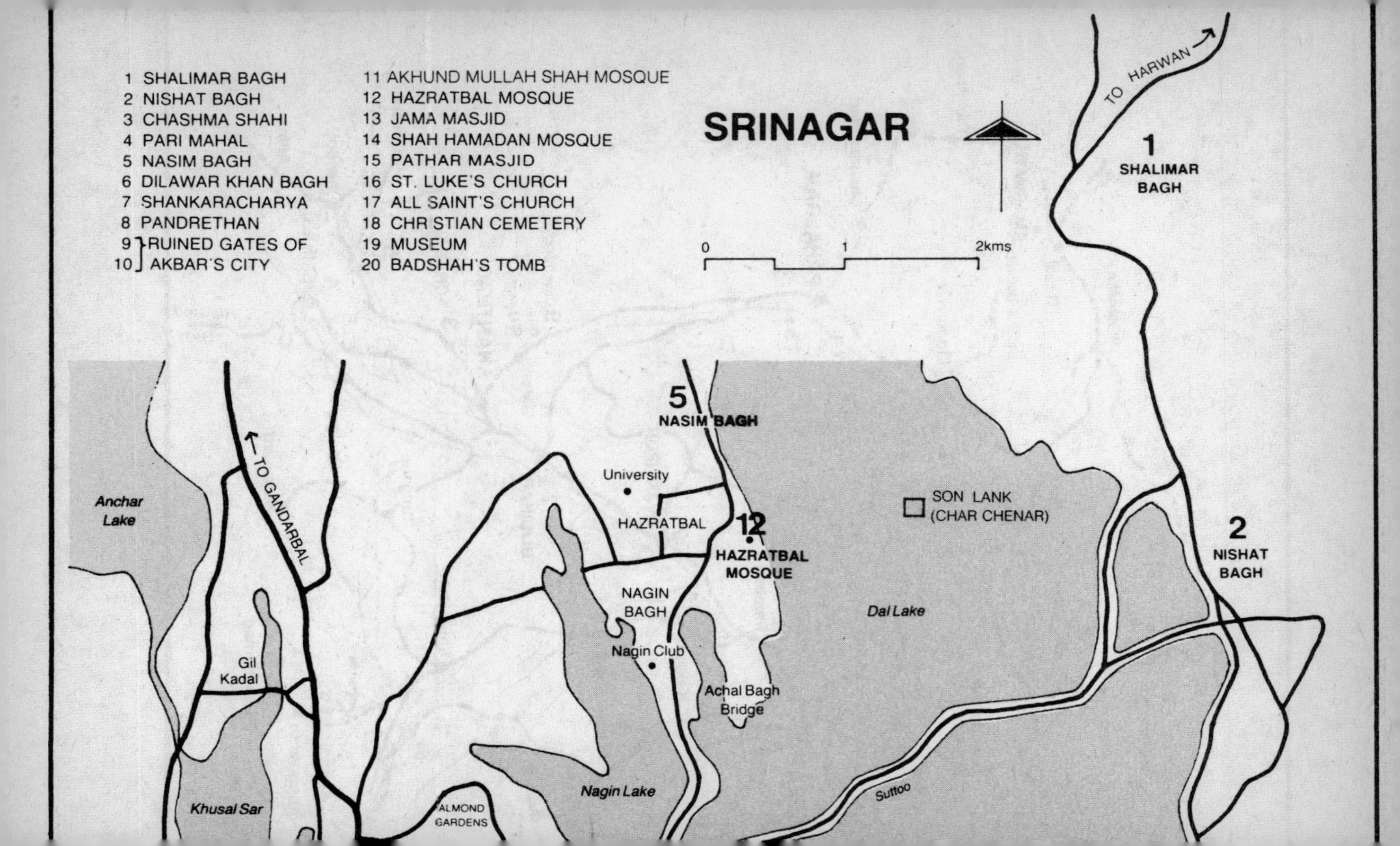
SRINAGAR
1 SHALIMAR BAGH
2 NISHAT BAGH
3 CHASHMA SHAHI
4 PARI MAHAL
5 NASIM BAGH
6 DILAWAR KHAN BAGH
7 SHANKARACHARYA
8 PANDRETHAN
9 RUINED GATES OF
10 AKBAR'S CITY
11 AKHUND MULLAH SHAH MOSQUE
12 HAZRATBAL MOSQUE
13 JAMA MASJID
14 SHAH HAMADAN MOSQUE
15 PATHAR MASJID
16 ST. LUKE'S CHURCH
17 ALL SAINT'S CHURCH
18 CHR STIAN CEMETERY
19 MUSEUM
20 BADSHAH'S TOMB
0
1
2kms
TO HARWAN
1 SHALIMAR BAGH
2 NISHAT BAGH
5 NASIM BAGH
University
HAZRATBAL
12 HAZRATBAL MOSQUE
SON LANK (CHAR CHENAR)
Dal Lake
NAGIN BAGH
Nagin Club
Achal Bagh Bridge
Nagin Lake
Suttoo
ALMOND GARDENS
TO GANDARBAL
Anchar Lake
Gil Kadal
Khusal Sar

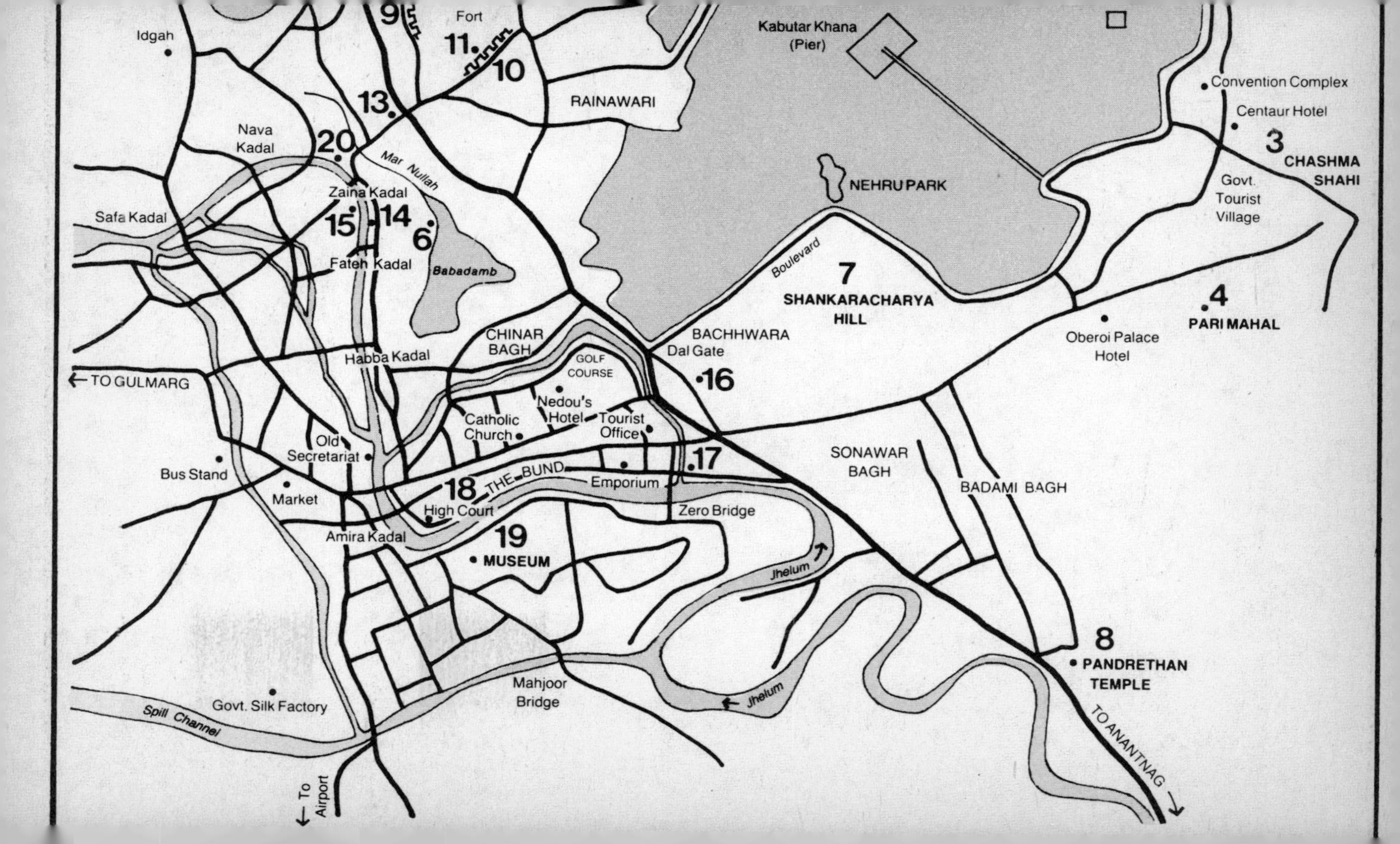
Idgah
9
Fort
11
10
Kabutar Khana (Pier)
Convention Complex
Centaur Hotel
3
CHASHMA SHAHI
Govt. Tourist Village
RAINAWARI
13
Nava Kadal
20
Mar Nullah
NEHRU PARK
Zaina Kadal
15
14
6
Safa Kadal
Fateh Kadal
Babadamb
Boulevard
7
SHANKARACHARYA HILL
4
PARI MAHAL
Oberoi Palace Hotel
CHINAR BAGH
BACHHWARA
Dal Gate
16
Habba Kadal
GOLF COURSE
TO GULMARG
Nedou's Hotel
Tourist Office
Catholic Church
Old Secretariat
17
SONAWAR BAGH
Bus Stand
18
THE BUND
Emporium
BADAMI BAGH
Market
High Court
Zero Bridge
Amira Kadal
19
MUSEUM
Jhelum
8
PANDRETHAN TEMPLE
Mahjoor Bridge
Jhelum
Govt. Silk Factory
Spill Channel
TO ANANTNAG
To Airport

A Personal Introduction

A Personal Introduction

There were two main reasons for wanting to spend some time in Kashmir. One was that I collect old Kashmir painted papier mâché—things like boxes and candlesticks and trays—and thought I should find out something about it, and possibly write a booklet on the subject; and the other was that I was born in India, one of the last daughters of the British raj, and I longed to rediscover it with my husband and children. For various reasons we could only take our holiday in June, and since that is the hottest time of year in India and our children were small, Kashmir—which is pretty well perfect at that time of year—seemed the most sensible destination.

We began our holiday in an inexpensive hotel that we had picked with a pin from a list sent to us by the Indian tourist office. This was mostly because of my mother's dire warnings about houseboats in Kashmir. Apart from obvious dangers such as falling overboard and drowning, it was well-known, she said, that the washing-up was done in the same lake water into which the lavatories drained. Indeed, she had taken me to Kashmir when I was a baby and we had stayed on a houseboat and I had contracted dysentery and was only saved by a rapid move off the boat and up to the Kashmiri hill station of Gulmarg. . . .

The hotel we picked turned out to be the staging post for Indian coach parties which arrived and left in the early hours of every day. In my experience Indians are the most excitable and noisy travellers in the world, and sleep was impossible as they yelled and shouted to each other at three and four in the mornings, dragged bed-rolls and luggage across the floors above and below our rooms, and loaded the buses outside our window with more cries and bumps and thuds as they hurled suitcases up to be stacked in the giant roof-racks. Even the helpfulness of the staff of the hotel could not compensate for the disturbed nights, and after a week we decided to forget about my mother and look for a quiet clean houseboat (where we would, of course, check on the washing-up water). By chance we found the 'Mehboob', a beautiful new-ish boat moored next to a tiny garden in the Dal lake. It had three vast bedrooms with bathrooms *ensuite*, a dining room with heavy, elaborately carved Kashmiri furniture, a drawing-room with chandeliers and huge square armchairs, and a lacey carved wooden verandah on which to sit and watch the world paddle past. We were lucky—the 'Mehboob' was only empty because there had been riots in Kashmir that spring, and many tourists and travel companies had cancelled their Kashmir bookings.

A *shikara*, the oriental gondola, was sent by our houseboat owner to fetch us to our new home. It was called the 'Galloping Snail' and was to be ours for the holiday. We were no longer at the mercy of the boat boys and taxi and tonga drivers who had lurked outside the hotel. Actually, we had already made a bid for freedom by hiring bicycles in the bazaar, but our sightseeing was frustrated by not being able to find any sort of child seats to attach to the back.

Now in the 'Snail', lying back on cushions under a flower-patterned awning, we made excursions to the Mughal gardens of Shalimar and Nishat. We explored the waterways of the ancient tumbledown city of Srinagar and took a trip up the Jhelum river—for which extra boatmen had to be hired as the current against us was so strong. One day the boat took us to the foot of the Hari Parbat hill, and from there we walked up to look at the fort that crowns the summit. We were not allowed in, for we had not done our homework properly, and had no written permission from the tourist office. Another afternoon we visited the mosque at Hazratbal and then picnicked on the little Char Chenar island nearby—once a favourite Mughal pleasure spot. Sometimes we abandoned the 'Snail' and climbed nearby hills instead, and once we walked along the causeway, or *suttoo*, as the locals call it, which spans the Dal lake from the suburb of Srinagar called Rainawari to the Nishat garden on the other side. (We had managed to persuade a taxi to take us to the Rainawari end of the causeway and collect us again from the Nishat end.) The walk is about four miles long and passes through an enchanted world of pretty villages, green willows and dappled waters—it is a glimpse of a Kashmiri way of life that you would not normally see. The 'Snail' ferried us to the lovely Nagin lake which many visitors prefer to the Dal because it is so quiet and tranquil, though personally I enjoy the bustle of life on the Dal lake. But most often we simply pottered around in the 'Snail' among the water lilies and incredible pink lotuses and the floating gardens.

The floating gardens are an ingenious way of creating more land—indeed so much new land is being created that the authorities claim the lakes are under threat. What happens is this: the farmer ties a whole lot of lotus roots together to make a raft. Lotus roots float, so the raft is buoyant and remains on the surface of the water. The farmer puts soil on the top and plants his vegetables which flourish in their water-borne bed, and the only problem he has is that when the crop is ready to harvest someone may cut the floating garden from its mooring and tow it away.

It has to be said that on these expeditions our privacy and peace were constantly interrupted. Kashmiris are the most persistent and thick-skinned salesmen, and as our little boat skimmed across the waters of the lake, vendors on similar craft would slip silently out from the shadows where they had been lying in wait and follow us like sinister crocodiles. As E. F. Knight, a *Times* correspondent writing in 1893, said, 'All this may be amusing at first, but one soon wearies of it. One discovers that to enjoy peace one must be a trifle brutal. Some sahibs obtain comparative privacy with a stick...'. My own advice is to be firm. Say 'no' and look away—the moment you falter, even by the flicker of an eyelid in the direction of the goods being sold, you are lost.

Other less aggressive floating shops came to our verandah in the evenings: Mr Wonderful, who paddled a boat piled high with flowers—roses, pinks, marigolds, irises, delphiniums, lupins; a groçery boat that sold dusty tins of provisions, faded packets of Aspirin, Elastoplast and Tampax, Indian whisky and (offered in whispers) hashish. And then there was Lassa Mir, the tailor, who was always trying to persuade me to order some Kashmiri-style drawstring trousers. 'But Mr Mir', I protested feebly in the end, 'I am afraid they might bag at the knees.' 'No, no, Madame', he replied vigorously, 'I assure you they will be cut just like my own.' He stood up to demonstrate, and it was even worse than I feared—though his own legs were straight, his trousers were still kneeling, but after that I hadn't the heart to refuse, though, apart from their famous fine woollen shawls, clothes were really the last things I wanted to buy in Kashmir, where the chain-stitched rugs known as *numdahs*, the beautifully carved wood, the fine hand-knotted or embroidered carpets and the prettily painted papier mâché are all unique and good value.

In 1894 an English memsahib called Mrs Burrows travelled to Kashmir with her husband and baby and wrote a charming little guide book called *Cashmir en Famille* in which she suggested that since 'nearly everyone will want to order some of the fascinating Cashmir work' it was sensible to budget for it when planning the holidays. She herself spent Rs 400 on 'works of art', which was only slightly less than all her servants' wages for the six-month trip. Mrs Burrows also suggested that the best plan was to order one's 'art' as soon as possible after arrival 'so that things may be finished and examined before the visitor leaves Cashmir'. Nowadays shortage of time makes it impossible for most tourists to order special things, and for them the best advice is exactly the opposite—look around at what all the dealers have and decide towards the *end* of the holiday what the best buys are.

Apart from the state-run handicrafts centre, based in the house that was once the British Residency in Kashmir, there are excellent merchants all over the city. Suffering Moses on the Bund is the best known, but he is a most eccentric and unpredictable old man, and at the precise moment you set your heart on something in his shop he usually decides that he cannot bear to part with it and refuses to sell it to you. 'Insufferable Moses' we came to call him, though with some affection. Often, maddeningly, and for no particular reason, he chooses not to open his shop at all and then sticks an unconvincing notice on the door saying CLOSED FOR STOCKTAKING. The present-day Suffering Moses is, apparently, the third dealer to bear the name—which was first coined by an English visitor in the last century for a merchant called Moses. The story goes that the two men were bargaining over some beautiful object, and, as he grudgingly lowered his price rupee by rupee, the Kashmiri dealer looked so forlorn and miserable that the Englishman said 'Oh how you are *suffering*, Moses', and Suffering Moses he was for the rest of his days.

A favourite merchant among the British in the old days was a tailor called Butterfly who was skilled at making fine lingerie. Before the war, memsahibs all over India would send him their underclothes to copy—until one year, to their horror and embarrassment, Butterfly brought out a catalogue in which all was revealed—there was their underwear, sketched in detail, priced, and with their names on them—Lady Snook's French Cami-knickers, Rs 25, etc.

Kashmir is one of the many places in the world which could easily have been the inspiration for the words of the hymn 'Every aspect pleases and only man is vile'. Man has contributed precious little to the sum of beauty in the valley and is responsible for all the unpleasant things, such as traffic, exhaust fumes, concrete buildings, noise, litter, and so on—not to mention corrugated iron, which has understandably, but sadly, replaced the old earthen roofs of the houses on which not so very long ago grass grew and flowers bloomed.

Our houseboat owner had various suggestions for excursions to beauty spots away from the bustle of Srinagar, so we took taxis—which he booked for us so that we would not be taken for a ride in more ways than one—to the Wular lake, which is enormous, wild and desolate, and to Manasbal lake, where we were excited to find a tiny Hindu temple dating back 1000 years partially submerged in the waters of the lake below the Government Rest House and garden. We drove to Sonamarg and to

Yusmarg—peaceful green places in the hills at opposite sides of Kashmir, and we visited Gulmarg, scene of my recovery from dysentery as a baby. I doubt whether my mother would recognize Gulmarg now. To get there in her day you had to walk or ride a pony up the last few miles of mountain from Tangmarg, and wretchedly poor porters used to carry up the pianos and potted palms that the sahibs thought indispensable in their holiday houses for few miserable rupees. Now a proper road goes all the way to Gulmarg, and the place is a busy ski resort in winter and a favourite with day trippers from Srinagar in the summer. We were puzzled that so many of them were carrying enormous thermos-flasks, and noticed, too, that everyone seemed to be keyed up to a higher pitch of excitement than is usual, even for Indian travellers. The cause of this became apparent when we reached Khilanmarg, yet another beauty spot a stiff walk or pony ride up the hill from Gulmarg. There the snow-line starts and several of the Indian tourists who had never seen snow before went wild with excitement, jumping in it and rolling, starting snowball fights and stuffing their thermos-flasks full of snow to take down to show the folks back home.

But the best way to escape from the rest of humanity in Kashmir is by heading for the hills to do some trekking. Hotels or houseboat owners make the arrangements, and the trek can be as tough or as tame, as luxurious or as frugal as you specify. My husband took himself off for an energetic four-day walk in the Liddar valley—this came into the frugal category for he was alone, apart from a guide (who also put up his tent, lit a fire and heated a can of beans every night) and a pony that carried the tent and tins. Even out there on the edge of the Kolahoi glacier, miles from anywhere, he met a group of elderly New Zealanders doing the same as him, but in that great emptiness he was, he confessed, rather pleased to see them. At the end of his trip I went to meet him at a prearranged rendezvous and came upon a scene of blissful serenity—in a lovely, empty valley I found my husband reading in the shade of a tree beside a rushing river while his guide dozed contentedly and his pony grazed nearby.

On our way back to Srinagar, in yet another taxi, we decided to make some detours in order to look at the great ruined Hindu temples at Martand and Avantipur, and the Mughal gardens at Achabal and Verinag. We had no idea, as we explored these sites, that we were following the exact route taken 300 years before by François Bernier, one of the first European travellers in Kashmir. How much more we would have enjoyed ourselves if we had only read his lively descriptions of the country—and the things we were seeing—in those far-off days when Kashmir was famed as the Paradise of the Indies, and its people admired as far superior in culture and talent to anyone else in the subcontinent.

But we had never even *heard* of Bernier, nor about the other early travellers from Europe, all of whom had fascinating tales to tell. In fact, our ignorance about Kashmir in general handicapped us throughout our six-week holiday, for while the natural beauty of the valley is spectacularly obvious, its past is complex and not easy to unravel on the spot. To sightsee successfully, rather than just admire the scenery, you must know something of the historical background and possess the instinct of a detective. For instance, at the famed Shalimar garden my husband was browsing round inside the black marble pavilion built by the Mughal emperor Shah Jahan, when he spotted a small section of the original painting that must once have covered the walls. There is only a small patch of this exquisite, delicate work—in gold leaf and black and yellow and red—left. The decoration on the rest of the walls is modern and much cruder. We were rather excited by our find and still more so, later on, when we came across François Bernier's description of the pavilion in 1665, finding it tallied with what we had seen. 'The whole of the interior is painted and gilt', he wrote, 'and on the walls of all the chambers are inscribed certain sentences, written in large and beautiful Persian characters. (Among others the celebrated legend "If there be a Paradise on earth, it is this, it is this").' Again, it was only by chance, on our walk down Hari Parbat hill, that we came upon the small but exquisite grey stone mosque that was built by the Mughal prince Dara Shikoh, for it is not in any guide-book that I have come across. Neither have I found any references to the remnants of Nagar-Nagar, the town built by Akbar at the foot of Hari Parbat hill (below the grey mosque), though two huge gateways that were entrances to it still stand. Nur Jahan's mosque in Srinagar city, the Pathar Masjid, usually gets only a passing mention, and the garden of Pari Mahal—which must be one of the most beautiful sites in Kashmir—in spite of the insensitive restoration work and hideous modern lamps which have recently sprung up all over it—receives so much less attention than the Shalimar garden, that it could be overlooked. Having said all that, I must confess that we had a great deal of fun and interest exploring neglected sites and digging out information for ourselves.

We did have one gem of a book lent to us by our houseboat owner. This was Dr Duke's *Kashmir Handbook*, written in 1888—ours was the revised edition of 1903. This guide gave us a feel for the place all right, but in the wrong century. Travellers to Kashmir in those days were a different breed. You have only to read the list of items that the doctor advises visitors to bring with them to see what a stoic and self-sufficient lot they were. Apart from medicines to counteract diarrhoea, dysentery, constipation, fevers, headache, snow-blindness, chilblains, frostbite, wounds and dog bite, and not forgetting that 'castor oil should always find a place', the doctor insists upon gum lancets, abscess lancets and some soft catheters. 'I mention the latter', he writes, 'as in 1875 had one of these instruments been available, a life might have been saved in the wilds of Kashmir. Since then I have always carried one in my gun case.' He then goes on to give instructions for un-dislocating your own shoulder or elbow:

> Get one shikari (No. 1) to hold the chest firmly with his arms. Get another man (No. 2) to put his knee under armpit from behind; make another (No. 3) extend the arm full length, raising it until there is freedom from all pain. Then tell him (No. 3) to pull upwards firmly: the chest man (No. 1) counter-extending. Try different angles forwards and backwards until the bone will go back with a jump. The pain at first is sickening. . . .

Dr Duke was a mine of rather gruesome information—it was

a bit like having my mother on board. There was not a sad way-side grave on the path into Kashmir that escaped his attention: one for a young Englishman who had drowned in a river, another one for a cholera victim, and a tiny one for poor little Harry, the five-month-old son of Lt W. E. Nuthall, who died and was buried at Naoshera. His parents added a postscript to the inscription on his grave: 'Travellers will do a kind act by instructing the Takidar to keep this grave in order.' One can't help wondering what has become of the grave in the hundred years since they wrote those words, so confidently assuming that there *would* always be responsible English travellers in Naoshera, and Takidars ready to do their bidding.

According to Dr Duke, the construction of the first proper road between India and Kashmir in the 1880's claimed many lives: fifty-four men died falling over precipices and twenty from snake bites. And sudden death on the road continued even after it was completed, for far too many people, Dr Duke writes sternly, harnessed their vehicles to over-fresh young ponies not trained to go downhill, and accidents were commonplace. On one occasion an English lady and gentleman were travelling in a tonga that took a corner too fast and went hurtling off the road and over the edge of the cliff. The lady was saved by the presence of mind of the man who, as they toppled, cried out 'JUMP'. 'With her umbrella in one hand she gave a spring, fortunately clearing the side and escaping certain death', writes Dr Duke, and goes on to relate how the lady's presence of mind in taking her brolly then saved them both from another death from sunstroke as they waited on the side of the road to be rescued.

Dr Duke is particularly eloquent on the nightmarish subject of rope bridges—things that the modern traveller to the valley of Kashmir can be truly thankful that they will not encounter. The most notoriously hair-raising rope bridge in Dr Duke's days was at Uri, near the new road from the rest of India. 'It is curious (I myself have witnessed it)', he wrote, 'that some of the boldest cragsmen, men who can face the most ghastly precipices have an aversion to crossing this bridge.... Nervous people who must cross generally have their eyes bandaged and are conveyed over pick-a-back or tied in a kilta [a porter's basket] and carried as a load.'

The doctor is always practical—in the middle of a description of some Mughal ruins he breaks off to say that there are, nearby, 'several fine Rhododendrons from which cuttings may be taken'—but he has an engagingly humorous side as well. Once, staying in a *dak* bungalow (a government rest-house) he flipped back through the visitor's book and came across an amusing exchange. Two vicars had stayed in the bungalow and had written in the book: 'Service good; but bugs and flease in abundance.' The next few guests made no comment on this, but then a Captain X, obviously a bit of a wag, spent the night and wrote 'Quite satisfied. The Reverend gentlemen above took all the bugs and flease away with them.'

Much as we loved Dr Duke, though, he was more of a curiosity than the companion we would have chosen for our travels. However, I now turned to my reason for being there, the art of papier mâché painting. I talked to master craftsmen and watched them at work with their brushes made of cats' hairs and their broken cups and saucers full of paint. I looked at the fine pieces of papier mâché in the dusty National Museum, and, with the help of a Kashmiri friend, sought out obscure Imambaras (Muslim halls for theological discussion) with painted papier mâché ceilings. I also inveigled myself into the elaborately decorated Old Secretariat, which used to be called the Shergarhi, and was the palace of the maharajas—which presumably explains why, not so long ago, a priceless horde of jewellery was discovered hidden away in there. (The descendants of the maharajas of Kashmir and the Indian government are still involved in a legal battle as to the ownership of the treasure.) And then one limpid watery dawn, the 'Galloping Snail' took us across the lake for the last time to catch the taxi to the airport.

Back in Europe I began writing my booklet on papier mâché, but I found it impossible to stop at that, or even at the arts and crafts of Kashmir. All sorts of people were jostling to get into the story and needed to be included as well: the Mughals, the Afghans, the Persians, the Sikhs, the British, the Indians—they have all had a hand in Kashmir. And then there was the great king of Kashmir who introduced the arts into his country in the first place, not to mention the intrepid early European travellers who brought the news of this other Eden back to the West, and the romantic Irish poet who never visited Kashmir but glamorized it so impossibly in his poem 'Lalla Rookh'. . . .

I decided to include them all, to abandon my first idea and instead write a *general* book on Kashmir, a hotch-potch of all that I had wanted to know when I was there but only found out on my return. I present it humbly, for I am not an historian, nor a geographer, nor a botanist. I make no claims for the pages that follow—only that I would have liked, very much, to have had them in my hands when I first explored Kashmir, that garden of so many secrets.

CHAPTER ONE

The Secret Garden: An Informal History

Who has not heard of the vale of Cashmere,
With its roses the brightest that earth ever gave,
Its temples, and grottos, and fountains as clear
As the love-lighted eyes that hang over their wave.

—from 'Lalla Rookh' by Thomas Moore

CHAPTER ONE

The Secret Garden: An Informal History

I do not believe that it is possible to write a single original sentence in praise of Kashmir—so much has already been said, and by such an extraordinary variety of people. Mughal emperors have flattered it in their most sugary prose, pilgrims and priests have softened at the sight of it, solemn British civil servants have waxed lyrical about the changing light on the mountains or sunset over the limpid lakes, scientists have turned into poets on the subject of Kashmir, worldly travellers have had to apologize for their child-like enthusiasm, and memsahibs have polished all the clichés about emerald pastures, sapphire waters and pearly snows.

Kashmir has been compared to Switzerland, the Italian lakes, the English lake district; to France, Greece, the Austrian Tyrol, and most often of all to the Garden of Eden itself. But good looks are obviously as troublesome a burden for a place as they can be for a person, and the story of Kashmir is rather like some lurid Victorian novel in which the heroine's rare beauty brings her nothing but disappointment and unhappiness. Kashmir's reputation as the nearest place to paradise that earth can provide has not made the little state prosperous nor its people happy—far from it, it seems to have condemned them both, through history, to much misery and suffering.

At the present time the geographical area that constitutes Kashmir is divided between India and Pakistan. Most of Kashmir lies within the political boundaries of India, with a portion to the west having been taken over by Pakistan. This unhappy state of affairs goes back to the troubled days of 1947 when predominantly Muslim Pakistan painfully separated itself from predominantly Hindu India. Kashmir then found herself uncomfortably sandwiched between the two countries, with both insisting that she should belong to them—Pakistan because the great majority of Kashmiris were (and still are) Muslims, and India because the ruler of Kashmir, Maharaja Hari Singh, was a Hindu.

It was agreed around the time of Partition that the only fair way to solve the dispute was to let the people of Kashmir decide for themselves whose side to join, but before any referendum could be organized an accident of fate forced history's hand—the country was suddenly invaded by an unofficial army of about 2000 Pathan tribesmen from Pakistan who made inroads into Kashmir frighteningly quickly, looting, raping and burning as they went. The Maharaja of Kashmir, panic-stricken, appealed to India for help, and was reassured that troops would be sent up at once. In return, Kashmir was to ally itself to India. This agreed, Indian soldiers duly arrived and drove out the marauding tribesmen. These Pathan invaders have been described as 'valiant warriors' by a Pakistani writer and as 'bloodthirsty raiders' by an Indian one.

India accused Pakistan of encouraging, if not actively supporting, the invaders; Pakistan denied any involvement, and protested furiously at the deal struck with the maharaja which handed Kashmir to India before the will of the people was known. War broke out between the two countries and continued for a year until, in 1949, the UN Security Council persuaded both sides to stop fighting and—for want of any better idea—the country was divided between them along the cease-fire line.

Again, it was suggested that the fair way to decide Kashmir's future was to let her people vote on it themselves, but India has always disagreed with the demands for a plebiscite and argued from different premises, and so, to this day, there are two Kashmirs. Pakistan controls the rugged, mountainous, empty area of the north and calls it Azad Kashmir (Free Kashmir), while India has the gentler valley of Kashmir, the heartland, containing the ancient capital, Srinagar—the part that has always been considered the true Kashmir, and which is the subject of this book.

It could be argued that neither India nor Pakistan has any convincing claim to Kashmir—and if it were a bleak and unattractive little country maybe neither of them would bother with it, but, as Shakespeare pointed out long ago, 'beauty provoketh thieves sooner than gold'. Being an earthly paradise brings its own problems.

Like a little apostrophe mark above the Indian subcontinent, the valley of Kashmir lies about 400 miles north of Delhi as the crow flies. Eighty-four miles long and twenty-five miles wide, it is about the size of half of Holland—which is, improbably, yet another of the countries with which it has been compared. What makes the place remarkable is that within this small area the land rises 13,000 feet—from the lush valley floor, which is already high at 5000 feet, to the snow-capped peaks of the encircling mountains which average 18,000 feet. It is the extraordinary changes of scenery in this rapid climb from temperate to tundra which make it possible to compare Kashmir with all those other contradictory countries. One Victorian traveller gazed at these layers of landscape and wrote that nature could only have achieved Kashmir in collaboration with an artist. Another summed it up more practically—'every hundred feet of its elevation brings some new phase of climate and vegetation, and in a short ride of thirty miles one can pass from overpowering heat to a climate delightfully cool, or can escape from wearisome wet weather to a dry and sunny atmosphere.'

The floor of the valley is flat, a lush patchwork of different shades of green, glinting with waterways and glassy lakes that mirror the poplars and willows around them so precisely it can be a puzzle to know which way up to hold photographs. Next come gentle, rounded foothills where, in spring and summer, cherry and almond trees, wild roses, irises, lupins and whole botany books full of other plants bloom as extravagantly as though they were cultivated. Then the slopes become steeper and are taken over by dark pine forests where the ground is soft and springy with layers of fallen needles, tumbling streams rush and clatter over stones and wild violets grow in hidden places. Further up it becomes too high for trees, the forests peter out, only stubby grass grows between the rocks, and the snowline is in sight. From this height you may be able to glimpse beyond the mountains that enclose the valley, and across the vast Himalayas where range after range of white peaks rise in every direction as far as the eye can see like petrified waves in a surrealist sea.

Frederick Drew, an English geologist working for the Maharaja of Kashmir in the last century, described these ranges as so desolate and bleak

> that it requires a considerable effort of mind of anyone who has never seen the like to picture to itself such a state . . . stretches of snow-field, wastes of stones, or else hillsides that bear forest untenanted by man, these occupy the chief space. . . . The one exception is Kashmir, which, set in the midst of the mountains, exhibits a fertile expanse, inhabited by an industrious people.

Kashmir, a small green footprint in this vast, hostile and colourless landscape naturally became known as more than just a pretty place. It has always been thought of as a land of milk and honey, an oasis, a coveted prize—and over the centuries greedy eyes have been turned on the valley by most of her less fortunate neighbours.

The curious fact that the valley is placed at the exact point, the *only* point, where, until recently, the three great religions of the East—Buddhism, Hinduism and Islam—met, has not enhanced her stability and security either. As Frederick Drew pointed out emphatically—'from Kashmir AND FROM NO OTHER SPOT IN ASIA one may go westward through countries entirely Muhammadan, as far as Constantinople; eastward among none but Buddhists to China; and southward over lands where the Hindu religion prevails, to the extremity of the Indian peninsula.' This, needless to say, has not been a very healthy crossroads and, inevitably, the people of Kashmir have been forced to adopt all three religions in turn—Hinduism, Buddhism, then Hinduism again, and then the Muslim faith.

Kashmiris are not an aggressive people; their character seems to have been shaped by nature's generosity to them—their land was fertile and productive and they had no need to seek new pastures or expand their frontiers. With one or two exceptions, Kashmir bred no great men of war. On the contrary, it was they who were repeatedly invaded by envious neighbours (though some of her own kings were so weak and depraved that this must occasionally have been quite a relief to the people). Over the centuries, the Kashmiris have had to learn the hard way how to survive conquest, oppression and extortion. They have had to teach themselves to be submissive, to bend with the wind, to evade all confrontation and to make the best of often appalling circumstances. As a result they have been castigated through much of history, by their masters as well as others, as a devious, dishonest and untrustworthy race.

When, for instance, Akbar conquered Kashmir in 1586 his court historian wrote: 'The bane of this country is its people'—and, throughout the 400 years since, visitors to the country have echoed his words. In 1904 Margaret Morrison, a normally pleasant, tolerant British memsahib, wrote scathingly:

> Though a fine race physically, they are for the most part a grasping, thievish, cowardly set. No quotation has so often run in my head since coming into the country as the opening lines of Browning's 'Childe Roland': 'My first thought was, he lied in every word.'

And Cecil Tyndale-Biscoe, a missionary who spent many years in the area around the turn of this century, said regretfully that the word 'Kashmiri' had become shorthand for a whole list of vices:

> To call a man a 'Kashmiri' is a term of abuse, for it stands for a coward and a rogue, and much else of an unpleasant nature. For instance, when giving a servant a character, a man whom you are dismissing and could not possibly recommend, you write: 'This man is a good specimen of a Kashmiri.' Everyone understands that such a man is not fit for employment.

But the Kashmiris have never been criticized for lack of intelligence or wit or skill—they may be the butt of all sorts of jokes and sayings but no one has ever suggested that they were stupid. Even Akbar had to admit that though the Kashmiri character might be seriously flawed, their skill at making things and their fine craftsmanship were outstanding and 'might be deservedly employed in the greatest cities'; and François Bernier, the French traveller who visited Kashmir in 1665, described the people as 'celebrated for their wit and considered much more intelligent and ingenious than the Indians.'

Bearing this out is the fact that the tiny area of Kashmir boasts one of the earliest written histories in Asia. Two hundred years before Chaucer and more than four hundred before Shakespeare (though several centuries *after* the great Indian Sanskrit epics and drama), Kalhana, a Kashmiri poet, traced the history of his country back 4000 years from 1148, the date at which he was writing, into the mists of time when, so legends tell, the valley was filled by a lake in which lived a horrible demon which terrorized the whole area, until the Hindu god Vishnu came to the people's help. Vishnu struck the mountains, near the town of Baramula, with his trident, the rocks cracked open and the lake water rushed out, leaving a sea of mud in the valley. The demon then tried to hide, but the goddess Parvati dropped the Hari Parbat mountain on top of him which crushed him to death, and so the inhabitants of Kashmir were at last able to live in peace.

Kalhana clearly loved his country and described it charmingly: 'Learning, lofty houses, saffron, icy water and grapes; things that even in heaven are difficult to find are common there.' We know practically nothing else about him, except that his father was a minister to King Harsha, one of Kashmir's most turbulent rulers, but he emerges through the pages of his great chronicle—the *Rajatarangini* (River of Kings)—as an exceptionally likeable and wise man, a shrewd and humorous observer of the human race—indeed, a sort of Kashmiri version of Chaucer and Shakespeare. And like theirs, the stories he tells prove that whatever the century and wherever the place, greed, passion and ambition remain exactly the same.

Hundreds of years later, when the Mughals conquered Kashmir, Kalhana's history was one of their most unexpected and exciting discoveries—so important that the court historian wrote this: 'When the Imperial standards were for the first time borne aloft in this garden of perpetual spring, a book called the *Rajatarangini* written in the Sanskrit tongue containing an account of the princes of Kashmir during a period of some 4000 years, was presented to His Majesty. . . '. Akbar, though he could apparently neither read nor write, greatly admired learning and ordered that the *Rajatarangini* should be translated into Persian, the language of his court. All human life seems encompassed in the tale Kalhana unfolds, but among Kashmir's rulers weakness and corruption, vanity, cruelty or greed seem to have been the dominant characteristics, and down the centuries the story is mostly a grim one. In dark days Kalhana described the Kashmiri court as a place where only buffoons and jokers and 'whoever was well-versed in stories about courtesans' thrived, and in still worse ones he wrote: 'The royal assembly, filled with whores, villains, idiots, and corrupters of boys, was unfit to be visited by the wise.'

The poor in Kashmir always had an additional burden to endure, literally. For hundreds of years—until the end of the last century—there existed a system of forced porterage called *begar*. In the mountainous country of Kashmir there were no roads and therefore no vehicles. Human beings made cheaper beasts of burden than mules or donkeys, and so organized press-gangs would go out into the countryside brutally snatching able-bodied men from their land and families and forcing them to act as unpaid porters for whichever noble, courtier or merchant needed them. It was not uncommon for journeys to last for weeks, if not months, and the wretched victims would return home half-starved and broken in health, or they might die of cold or sickness on the way and never come back at all.

Only a handful of wise kings and benevolent rulers shine in Kalhana's history. The first of these—and he is also the first known historical name to appear in the *Rajatarangini*—is Ashoka, the great Buddhist king who ruled much of India in the third century BC. According to Kalhana, Ashoka built the first city of Srinagar at a place called Pandrethan which is on the outskirts of the present-day city. There are no buildings—not even ruins—left in Kashmir dating back to Ashoka's time, nor that of his son, who is supposed to have built the first temple on top of the Shankaracharya hill in Srinagar. In fact there are no traces of any of the dozens of Buddhist shrines that, by Kalhana's account, once graced the valley.

After Ashoka, Kashmir gradually reverted back to the Hindu religion, and between the sixth and tenth centuries, which was the heyday of Hinduism, many temples were built around the valley. The ruins of some of these are still there to see, and you have only to look carefully at the walls of houses and mosques—and particularly at the embankments of the Jhelum river in Srinagar—to find interesting blocks of cut stone that were obviously removed from others and re-cycled. Arthur Neve, a missionary doctor in Kashmir for many years at the end of the last century and the beginning of this one, made a particular study of these old Hindu temples in Kashmir and wrote: 'ancient India has nothing more worthy of its early civilization than the grand remains in Kashmir, which now feebly excite the wonder of European tourists . . .'. All the old temples were dedicated to the god Shiva and, apart from the large ones at Avantipur and Martand, which are more elaborate and set in enclosures with grand gateways and colonnades, all of them follow the same pattern. They are sturdy, solid, stone sanctuaries between six and ten feet square, with heavy pyramid-shaped roofs, and each one probably stood in a pool of water. Every sanctuary once contained a stone phallus or lingam which the faithful would have walked around in prayer, and garlanded with flowers. The remains of temples like these can be found all over the valley (including the half-submerged one in Manasbal lake mentioned in the Introduction), but the easiest to find and the best-preserved example is the little temple at Pandrethan on the edge of Srinagar. This is where Ashoka built his capital, but the temple is much later than that; it was probably built in the tenth century. Today the Indian military have taken over this area, but there is nothing to prevent people visiting the temple. It stands in a small park behind a rusting tank which was captured from the Pakistani army and has now become a war memorial on the left-hand side of the Srinagar/Pahalgam road. The temple sits in a pool of water which visitors cross on a stone walkway. This walkway doesn't seem to have existed in the last century, for then every British traveller who went to Pandrethan complained of the horror they felt at the thought of wading or swimming through the slimy black water to reach the temple. Arthur Neve recommended the Pandrethan temple as being 'well worth inspection', particularly the inside of its roof, which is 'covered with sculpture of such purely classic design that an uninitiated person who saw a copy of it on paper would at once take it for a sketch from a Greek or Roman original.'

Of all the temple ruins to be found in Kashmir today, the most dramatic date from the reign of the next great king after Ashoka. This was Lalitaditya-Muktapida who came to the throne a long time later, in AD 725. Lalitaditya was a soldier king who expanded the frontiers of Kashmir through the surrounding hill kingdoms and down to the plains of the Punjab. But when he was not marching he was building, and it is for his great temple at Martand that he is remembered. The temple at Martand was built on a plateau above the present-day town of Mattan which is on the road to Pahalgam. (Mattan itself has a spring and a large pool full of fish which are so used to being fed they swim after visitors walking around the edge. Like many natural springs, it has always been a holy place, and there are temples and shrines all around. Indeed, it is hard to believe that Mattan, so full of traffic and crowds, was once a quiet green place, highly recommended by British Sahibs as a camp site.)

Earthquakes, iconoclasts and vandals long ago reduced Martand to ruins, but nonetheless it is still most impressive, and in its prime it must have been breathtaking. The temple—which had a sanctuary, a choir and a nave—stood in a large quadrangle with eighty-four carved stone pillars making a graceful colonnade around it. The western entrance, a magnificent archway, was approached by a wide flight of steps, and there were various side chapels. Arthur Neve calculated that the massive temple roof cannot have been less than seventy-five foot high. Unfortunately, the rubble from the collapsed roof has tended to confuse the site. The temple's simplicity of outline, the proportions of its pillars and colonnade, the triangular pediments above the doorways, are all strikingly classical and must have been inspired by Greek architecture, via Alexander the Great's conquests in north-western India. Even its site on a high plateau is rather Parthenon-like. Sir Frances Younghusband, who was British Resident in Kashmir in the early years of this century, wrote: 'No temple was ever built on a finer site. It is one of the most heavenly spots on earth... there is about it a combination of massiveness and simplicity, and of solidity combined with grace, which has earned it fame for a thousand years.'

The next leader of stature, Avantivarman, came to the throne in AD 855. He was, by Kalhana's account, a gentle and humane man whose first act on becoming king was to give the previous ruler's ill-gotten treasure to the poor—'for who could delight in necklaces cursed and unholy, which have been torn from the necks of the dying.' Scholars, artists and poets were promised 'great fortunes and high honours' and consequently flocked back to court, and fine temples were built at the king's new town of Avantipur. The ruins of these temples are easy to see. There are two of them, called the Avantiswara and the Avantiswami temples, about one kilometre apart, beside the Srinagar/Pahalgam road, just before Avantipur. As at Martand, the temple buildings have been ravaged by time, earthquakes and vandals, but their gateways, paved quadrangles and general layout can be seen clearly, and there are magnificent examples of stone carving—including, it is said, a portrait of Avantivarman and his queen on a block of stone to the left of the Avantiswara temple.

The real star of this king's reign was an engineer called Suyya who organized a drainage system for the valley. Year after year there were devastating floods in the Kashmir valley, until Suyya cleared the rocks and silt that were clogging the outlet of the Jhelum river and the water drained away, leaving the land 'covered with mud and wriggling fishes'. Suyya built massive stone embankments to contain the Jhelum river and is even supposed to have moved the site of the confluence of the Jhelum and its tributary, the Sind, from the town of Parspor to Shadipur.

Kalhana wrote that even in his day you could still find 'old trees which bear the marks of the boat ropes fastened to them' on the banks of dried-up river beds. Suyya's reward was to be allowed to build his own town, and he chose a site for Suyyapore, as it was called, on the banks of the river he had tamed. This is the modern town of Sopor.

There were no more wise and benevolent rulers between Avantivarman and the close of Kalhana's chronicle in 1149, but there were lots of rather evil and intriguing ones, such as the tenth century Queen Didda, a sort of Kashmiri version of Lucrezia Borgia or Messalina, poisoning all her rivals, and handsome King Harsha who ruled in Kalhana's own time and who seems to have gone mad halfway through a promising reign, so that it ended in terror and bloodshed. 'How is it to be related', wrote Kalhana sadly, 'that story of King Harsha which has seen the rise of all enterprises and yet tells of all failures.'

For nearly two more centuries after Kalhana's death Kashmir tottered on under a succession of unworthy kings, and the only extraordinary part of the story is that in all those years no adventurer, no mercenary, and no foreign power seized control of the valley, for, in the world outside, turbulent and momentous changes were taking place. Invading Muslim armies from Afghanistan conquered and converted the whole of northern India, while much of the rest of the known world was overrun by Mongols led by Genghis Khan who poured out from their homelands in the steppes of Asia. Kashmir remained untouched—an island of Hinduism in a sea of alien religions, until 1319 when a Mongol raiding party swept down into the valley, looting and killing and capturing slaves. This was only a hit-and-run affair, but it was to be the catalyst that changed Kashmir's history, for the Hindu king ran away, leaving chaos in the plundered country. Into this vacuum stepped Rinchen, a popular Tibetan princeling who had come to Kashmir to seek his fortune. He was crowned king in 1320 and chose as his chief advisor a Muslim, Shah Mir, who had been in the valley for years, and whose wisdom and sharp wit had gained him a place even at the Hindu court. Rinchen's admiration for Shah Mir soon persuaded him to become a Muslim himself, and it was in this way, calmly, peacefully, by conviction and conversion rather than by bloody conquest, that Kashmir made the dramatic change from Hinduism to Islam.

After Rinchen's death, Shah Mir manoeuvred himself onto the throne and his descendants went on to rule the valley for more than 200 years. One of them, Sultan Zain-ul-Abidin, is still revered by Kashmiris today as the wisest king they ever had. Padshah or Badshah (meaning Great King) they call him, and they speak warmly about him as a familiar father figure—imagine the British still feeling sentimental about Henry VI, his contemporary in England.

Zain-ul-Abidin was nineteen when he came to the throne in 1420, and he seems to have been one of those characters that history throws up only too rarely—a man in whom wide interests and great talents were combined with a genuine love for his subjects.

It was the religious tolerance of the new sultan that first impressed his people, for his father had been an austere and fanatical Muslim, the man popularly believed to be the one responsible for the destruction of the Hindu temples at Martand and Avantipur. Badshah was sympathetic to Hindus, and employed many Hindu scholars at court, one of them being Jonaraja, an historian whom he commissioned to bring the *Rajatarangini* up-to-date. He also abolished the laws and taxes discriminating against Hindus that had been introduced in previous reigns.

Badshah was, as a Muslim, both devout and humble. When a greatly respected holy man and hermit, Nur-ud-Din, died, Badshah walked alongside the coffin to its last resting place in a town called Charari Sharif about twenty miles from Srinagar. Nur-ud-Din is looked upon as the patron saint of Kashmir, and this tomb is today a very popular place of pilgrimage.

Badshah built the Zaina Kadal, also known as the Fourth Bridge, which still exists in Srinagar, as well as a wooden palace for himself said to be so tall that 'it humbled the pride of the peaks of the Himalayas'. This has long since disappeared.

been the island in the Wular lake called the Zaina Lank. There was an ancient legend which told that long ago a settlement of people lived where the Wular lake is today. These people, rather like those of Sodom and Gomorrah, had sunk into depravity and sin, and a holy man warned them that they should mend their ways or the gods would punish them. Naturally they ignored him, and one day an earthquake shook the land, altering its contours so that rivers and lakes poured onto the evil town, drowning everyone in it. Hearing this legend (so the story goes), Zain-ul-Abidin sent divers down to the bottom of the Wular lake to investigate, and it seems that they found the ruins of a large stone temple and two golden statues. The statues were sold to pay for the great earthworks that followed, for Zain-ul-Abidin ordered an island to be built on top of the ruined temple. Whether he actually created it or not, this man-made island in the Wular lake, so quiet and isolated, certainly became one of the sultan's favourite places. He built a palace and a shrine on it and would retreat there to rest and meditate for weeks at a time. Later, in Mughal times, the Zaina Lank was a favourite picnic spot for the emperors and their courtiers, but they were the last people who cared about it and now Zain-ul-Abidin's buildings are in ruins. According to travellers in the last century there was a stone slab on the shrine that bore this inscription in Persian:

> May this edifice be as firm as the foundations of the heavens
> May it be the most renowned ornament of the Universe
> As long as the monarch Zain-i-Ibad holds festival therein
> May it be like the date of his own reign—happy!

Sadly, now this stone has disappeared.

Zain-ul-Abidin had a well-organized army and reconquered neighbouring Ladakh which Kashmir had lost shortly before he came to power. But dearer to his heart than any military expedition was the furthering of knowledge and the exchange of ideas and philosophies and skills. He sent ambassadors as far afield as Turkey, Mecca and Egypt, encouraging artists and craftsmen from many countries to come to Kashmir and practise their talents, until the place became a hive of industry, famous for its exquisite shawls, its paper, its delicate painted papier mâché, its superb wood carving and its fine embroideries and metal work. Descendants of these immigrant craftsmen are still living in Kashmir to this day, still creating the lovely things for which the valley became renowned.

Zain-ul-Abidin died in 1470 and was buried in a cemetery not far from his Zaina Kadal bridge. The grandest grave in the enclosure is a domed brick tomb decorated with blue-glazed tiles—this is Zain-ul-Abidin's mother's tomb. His own is nearby, inside an attractively carved stone enclosure with an imposing gateway which is believed to have originally surrounded a Hindu temple.

Most of what Zain-ul-Abidin had achieved in Kashmir was undone over the next few decades by his grandsons and great-grandsons, who squabbled futilely over the succession. One of them came to power no fewer than five times, and on each occasion was deposed by a cousin. Absorbed in their family feud, they were blind to the greater danger threatening them from a tribe called the Chaks. The Chaks were a tough, resilient people from the mountains beyond the valley who took advantage of the rivalry between the princes to gain power for themselves. They gradually infiltrated themselves into key positions, until the day came in 1555 when the leading Chak 'advisor' was able, literally, to lift the crown off the sultan's head and place it upon his own brother's.

But the year after the Chaks made themselves rulers of Kashmir a thirteen-year-old boy who was to be their undoing inherited the throne of the Mughal empire in India. This was Akbar, who rescued the valley of Kashmir from confusion and misrule, and established a golden age that has become a legend. Akbar was the first of three Mughal emperors who were to be closely involved with Kashmir.

Mughal is the Persian word for Mongol, and the Mughal emperors who ruled India were descended from those same Mongol hordes which had overwhelmed much of the world two centuries before. The immediate ancestors of the Mughals had settled down in Central Asia, in roughly the triangle made if you draw a line between the cities of Samarkand, Kabul, and Herat, and there, traditionally, the princes of the tribe battled with one another to carve out little kingdoms for themselves, endlessly attacking and capturing and losing each other's villages and strongholds. In 1526 Babur, one of these young princes, found himself so frustrated in his efforts to gain a bit of territory for himself that he decided to abandon the struggle and seek his fortune in the East instead. Babur and his followers fought their way into India, and he ended up in Agra, surrounded by a much bigger and better kingdom than any he could have established in his own homeland. Babur and his descendants—the Mughal dynasty—ruled India from Agra or Delhi for more than 300 years, brilliantly at first, but towards the end increasingly ineffectively. The very last Mughal emperor was deported to Burma by the British after the Indian mutiny in 1857.

Of all the Mughal emperors, Babur's grandson, Akbar, was the greatest. He expanded the frontiers of the kingdom to their widest limits, but also ruled wisely and with extraordinary tolerance and was an enthusiastic patron of scholarship and the arts.

At first, Akbar in India and the Chaks in Kashmir were careful with each other. They exchanged ambassadors, and the Chaks sent lavish placatory gifts down to the powerful young Mughal emperor, including on one occasion a Chak princess. (In fact they had to send her twice, for the first time she arrived Akbar, who was annoyed with the Chaks at that moment, sent her back.) But it was inevitable, given Akbar's desire to expand his kingdom and the Chaks' poor kingship and sometimes irritatingly arrogant behaviour, that eventually Akbar would attempt to add Kashmir to his empire. Accordingly, he despatched 40,000 horsemen and 20,000 foot soldiers to take the valley, and after tremendously fierce resistance, the victorious Mughal army entered Kashmir in 1586.

This was not, in fact, the very first time that any Mughal had set foot in Kashmir. A distant cousin of Akbar's, a dashing adventurer called Mirza Haider, had gone there to seek his fortune in 1542, and ended up becoming the power behind the sultan's throne. He was killed in a skirmish with the Chaks a few years later, and was buried in the same graveyard not far from the Zaina Kadal bridge where Zain-ul-Abidin and his mother lie. In 1824 an admiring British traveller, William Moorcroft, came across Mirza Haider's grave and, as a tribute to his fellow-adventurer, he had a fine headstone made for it, with an elaborately carved Persian inscription. The stone is still there, a little the worse for wear. It lies under some trees opposite the entrance to the graveyard.

Akbar did not travel up to inspect his new conquest for three years, but then he was delighted with the place. 'The country is enchanting', wrote Abul-Fazl, his court historian, 'and might be fittingly called a garden of perpetual spring.... Its streams are sweet to the taste, its waterfalls music to the ear, and its climate is invigorating. The flowers are enchanting and fill the heart with delight. Violets, the red rose and wild narcissus cover the plains...'. This was the start of a love-affair between the Mughal emperors and the Kashmir valley that lasted through the reigns of Akbar, his son, Jahangir, and his grandson Shah Jahan—and like all good relationships it benefited both sides and brought out the best in each.

The Mughals had come from high lands, and the flat, hot, dry, dusty plains of India made them homesick for mountains, rushing streams and invigorating air, and nostalgic for the flowers and trees of cooler climates. Soon after his conquest of Agra, Babur describes in his memoirs how he had gone on a fruitless search for a suitable site on which to make a garden near the Jamuna river: 'The whole was so ugly and detestable, that I repassed the river quite repulsed and disgusted. In consequence of the want of beauty and the disagreeable aspect of the country I gave up my intention...'.

Of course, in the end, the Mughal emperors *did* create marvellous gardens in the inhospitable Indian landscape, but it was difficult always working against nature, having to use wells and water-wheels worked by teams of buffaloes to produce every drop of moisture on which the gardens depended. What a contrast was Kashmir—so lush and beautiful that it hardly needed a man's hand to transform it into a pleasure garden. No wonder they christened the valley 'an earthly paradise'. That is exactly what it must have seemed, especially compared to what has been described as 'the earthly hell of the Punjab in summer'.

The Mughals and Kashmir were made for each other. The valley gave them a sanctuary from the unbearable heat of India, a resort in which they could regain their health, refresh their spirits and amuse themselves creating summerhouses and gardens. It could even provide all manner of skilled craftsmen and artists to help them realize their schemes. And Kashmir also supplied the royal larders in Agra and Delhi and Lahore with melons, apples, peaches, plums, apricots and grapes. 'Ever since the conquest of Kabul, Qandahar and Kashmir', reports Abul-Fazl, 'loads of fruits are imported; throughout the whole year the stores of the dealers are full, and the bazaars well supplied.' Cartloads of ice from Kashmir arrived daily in the capital—'all ranks use ice in summer, and nobles use it throughout the year'—and there were regular supplies of 'ducks, water fowl and certain vegetables' to supplement the royal diet. And of course Kashmir was the only supplier of saffron, that costly, delicate, coveted spice made from the threadlike stamens of a crocus that has, for centuries, been cultivated at Pampur, near Srinagar.

On their side, the Mughals gave Kashmir peace, stability, and a just administration, as well as the benefit of their extraordinary flair for architecture and garden design, and their enthusiasm and encouragement in all the decorative arts.

On a hugely grand scale the Mughals became like any family with a treasured holiday home. They returned eagerly whenever they could, longing to inspect the improvements and embellishments they had ordered on previous visits, delighted to see the place growing in beauty, forever thinking up something new to do. Jahangir's Kashmir diary, particularly, is full of the pleasure he felt at seeing his ideas carried out. When he inspected the garden and palace that he had ordered to be built at Verinag, for instance, he exclaimed enthusiastically that it had become 'a place such that travellers over the world can point out few like it.' Unlike any subsequent rulers of the valley, the

Mughals *gave* to Kashmir as well as taking from it, they enhanced its natural beauty and left it a more beautiful place than they found it.

The Kashmiris must have enjoyed the visits of the emperors and their exotic entourages—the arrival of the Mughal court was probably like the most spectacular circus on earth coming to town. Crowds gathered to see fabulously decorated elephants with glittering golden howdahs swaying on their backs in which the ladies of the harem sat hidden; the dashingly uniformed horsemen; the painted palanquins; the sleek courtiers in their silks and brocades; the elaborately caparisoned horses. The jewelled figure of the emperor himself would have been visible on a field throne—which Bernier, the French traveller, describes as 'a species of magnificent tabernacle with painted and gilt pillars and glass windows.'

To smooth the path for Akbar's first visit to Kashmir, 5000 labourers were sent ahead to build a road through the difficult mountain country that had to be crossed to reach the valley, but the journey from Lahore still took six weeks. Akbar finally entered Srinagar in triumph on 5 June 1589 and the royal standards of the Mughals were raised in Kashmir for the first time. He made his camp in a garden in the town that had been built by one of the Chaks, and almost his first action after arriving was typical of this tolerant and open-minded king who sought to reconcile all the religions and races he now led: he travelled to the temple at Martand (which was still standing then) and presented the Hindu priests with 'cows adorned with pearls and gold'.

Akbar spent a month exploring the valley and Abul-Fazl made notes on the flora and fauna of Kashmir, on the people and their habits and superstitions, and on the scenery. Reading them, it seems that some things in Kashmir have not changed at all, in particular life on the waterways. Then, as now, 'the people take their pleasure in skiffs upon the lakes'—and in the same kind of *shikaras* that skim so gracefully across the glassy surfaces of the lakes today. And when Abul-Fazl says that 'able artificers soon prepared river palaces' for Akbar, he could be describing the grander of today's houseboats.

Akbar continued his sightseeing in two further visits, in 1592 and 1597. He discovered that autumn in Kashmir was, if anything, more beautiful than spring. He saw the saffron fields in bloom and was enchanted: 'a sight that would entrance the most fastidious'. He travelled to see the gushing spring at Achabal: 'a delightful place for visitors, and a place of worship of the ancients. There is a limpid reservoir . . . occasionally a beautiful yellow spotted fish appears . . .'.

Not even the abundance of 'fleas, lice, gnats and flies', nor the horrors of Kashmiri music—'with each note they seem to dig their nails into your liver'—could spoil the pleasure he took in his new domain. He christened Kashmir his 'private garden', and directed his governor to begin work on a fortress that would be the Mughal base in the valley and ensure that it remained safely in their hands. An immense wall was constructed at the base of the Hari Parbat hill near Srinagar, and within it palaces, houses and gardens were built for Akbar and his wives and nobles. On the top of the hill was a fort (though not the one that crowns Hari Parbat today—that was put up by the Afghans at the beginning of the nineteenth century).

Two hundred Indian stone-masons were brought to Kashmir to work on Nagar-Nagar, as Akbar's stronghold was called, for though Kashmir had once produced magnificent workers in stone, as the battered ruins of Martand and Avantipur testify, the art seemed to have died out and by Akbar's time building in Kashmir was done in wood and brick. (The Jama Masjid and the Shah Hamadan mosque in Srinagar are good examples of the old wooden architecture of Kashmir. Both were built in the fourteenth century, both destroyed by fire several times and rebuilt in their present forms between 250 and 300 years ago. The Jama Masjid has no less than 300 magnificent wooden pillars holding up its curious pagoda-like roof, each pillar being the trunk of a giant deodar tree. The Shah Hamadan mosque, named in honour of a Persian mystic who came to live in Kashmir, is made entirely of wood, but the blocks are laid to look as though it is brick-built.)

No slave labour was used to build Akbar's fortress complex. There were set rates of reward for every job, and as Abul-Fazl records, 'a great many persons got their livelihood from the building of the fort. By means of the pay for their labour, they were brought out from the straits of want.' Nagar-Nagar was not completed until the reign of Akbar's son, Jahangir, and the most detailed description of it was written at that time by a Dutchman, Francis Pelsaert, who worked for the Dutch East India Company in Agra. Pelsaert probably never went to Kashmir himself, but based his description on other peoples' reports of the place.

> On the East side of the city lies a great stronghold with a wall of grey stone fully nine or ten feet thick, which joins it to a high rocky hill, with a large palace on the summit, and another somewhat lower or half way up, towards the North, as well as two or three residences with separate approaches.... In the centre of this fort is the King's palace, which is noteworthy rather for its elevation and extent than its magnificence. The Queen lives next to the King.... On the south-west live... other great nobles, all of whom reside within the fortress and round the hill....

Almost nothing is left of Nagar-Nagar today, apart from two imposing but crumbling gateways. One is to be found at the foot of the walkway up Hari Parbat hill, and the other, more difficult to locate, is not far from the Firdose cinema. In between and around these gates lie portions of crumbling masonry and chunks of cut stone that are probably relics of Nagar-Nagar and might be interesting to explore.

Akbar also laid out a great garden on the north-west side of the Dal lake which he called Nasim Bagh or Garden of Breezes. The site is still there, but today a busy road goes through the middle and most of the rest of it has become the grounds of a teaching college. The best view of Akbar's garden is probably from across the lake. From there the plantation of chenar (oriental plane) trees that once shaded his terraces and pavilions, stand out as a dramatic block of dark green. The small part of the garden that lies between the road and the lake is still pleasant, and in it there are the traces of old stone walls that must once have held up a terrace or been part of a building. Some houseboats are moored to this little patch of green. They can be rented, and Akbar's garden would be a romantic place to stay. Towards the end of the last century a *Times* newspaper correspondent, E. F. Knight, travelled through Kashmir and was invited by the British Resident to a picnic in Nasim Bagh. He was entranced:

> No more delicious spots can be found for open-air revelry than the fair gardens that surround the capital of Kashmir.... Indeed, the genius of picnic seems to rule the whole shore of the Dal.... I should not be surprised, by the way, if the very word picnic, whose origin I believe is unknown, were some old Kashmir name for the pleasant pastime of which this Happy Valley was the birth-place.

Nasim Bagh was the very first of the great Mughal gardens to be laid out on the edge of the Dal lake, but such was the Mughal passion for garden-making that hardly any time later, in Jahangir's reign, there were apparently no less than 777 gardens in the area. It is hardly surprising to learn that at one time the roses grown in Kashmir brought in a revenue of Rs 100,000 a year, nor that Jahangir's mother-in-law invented attar of roses—a scented essence made from the flower petals. Traces of these Mughal gardens can still be found around the Dal lake. When Constance Villiers Stuart was researching her classic book, *Gardens of the Great Mughals* (1913), she discovered, hidden in mud and refuse under the wooden platform of the old mosque at Hazratbal, two exquisite carved stone-fountain basins that had clearly once belonged to some grand Mughal garden. Modern travellers should not waste time searching for more *there*—a new mosque has been built at Hazratbal since Mrs Villiers Stuart's discovery.

Akbar's conquest of the Kashmir valley brought peace. The squabbles over the succession that had torn the country apart for so long were pointless now that it was ruled from India, and the Mughal presence was strong enough to discourage local rebels. Abul-Fazl wrote that under Akbar the valley became 'the secure and happy abode of many nationalities, including natives of Persia and Turkestan as well as of Kashmir'. Governors were appointed to administer the valley and these were kept on a tight rein. Governors could be punished for imposing heavy taxes, maltreating Hindus, or stirring up religious strife, and the people were not, it seems, afraid to complain about them if they felt they had cause. Jahangir sent this message to one of his errant governors, and though the language is flowery the threat is clear: 'Thy complainants are many, thy thanksgivers few, pour cloud water on the thirsty people or else relinquish thy administrative post'. The town of Srinagar was a prosperous place under the Mughals. Pelsaert, the Dutchman, says:

> The city is very extensive, and contains many mosques, as their churches are called. The houses are built of pine-wood, the interstices being filled with clay and their style is by no means contemptible. They look elegant, and fit for citizens rather than peasants and they are ventilated with handsome and artistic open-work, instead of windows or glass. They have roofs entirely covered with earth, on which the inhabitants often grow bulbs. . . .

But not all the power of the Mughal emperors could protect the Kashmiri people from more than their share of natural disasters. There was a terrible famine at the time of Akbar's last visit to the valley in 1597, and though grain was brought in from neighbouring states the situation remained desperate. Two Jesuit priests who accompanied Akbar (and whom we meet in the next chapter) described the pitiful sight of destitute mothers selling their children. In Jahangir's time there was an outbreak of plague in the valley, and when this was followed by a fire that devastated part of the city, burning 3000 houses, Jahangir wrote in his journal that he prayed for his people to be 'altogether freed from such calamities'. But they were not. In Shah Jahan's time famine struck again and the plight of the Kashmiris was so bad that it moved the nobles of the court to set up a relief fund for the valley.

Akbar died in 1605, having pushed the frontiers of the Mughal empire as far west as Kabul, as far east as Bengal, north to include Kashmir and south to the Deccan provinces of India. His son took the name Jahangir, or Holder of the World, at his accession.

As a prince Jahangir had visited Kashmir twice with his father and had fallen in love with the place. 'If one were to take to praise Kashmir', he wrote, 'whole books would have to be written . . . its pleasant meads and enchanting cascades are beyond all descriptions . . .'. As emperor he made many trips to the valley—between six and twelve—and once said that he would rather lose his whole kingdom than little Kashmir. Pelsaert was cynical about Jahangir's passion for Kashmir: 'The reason of the King's special preference for this country is that when the heat in India increases, his body burns like a furnace, owing to his consumption of excessively strong drink and opium. . . .' The Mughals did have a weakness for opium and alcohol, and Jahangir was no exception. He admitted that in his youth he regularly drank twenty cups of the strongest spirits a day, and was pleased with himself for cutting it down to six cups of alcohol and wine mixed. But it is clear from his diary that his fascination for Kashmir sprang as much from his great love of nature and his boundless curiosity about all wild things as from a desire to be cool.

It was Jahangir's custom when he travelled always to be accompanied by an artist, so that if anything particularly beautiful or unusually odd crossed his path it could be recorded immediately. Ustad Mansur was his favourite painter and in Kashmir he was kept busy: 'I ordered Ustad Mansur to draw its likeness,' Jahangir writes again and again; or, 'the flowers that are seen in the territories of Kashmir are beyond all calculation. Those that Ustad Mansur has painted are more than 100.' Even while relaxing and drinking his 'usual cups' in a lovely picnic place near a waterfall, Jahangir was alert and interested enough to spot an unusual bird diving into the water of the stream, order servants to capture it, examine it himself and then pass it on to Ustad Mansur for drawing. He watched the saffron being harvested at Pampore and found that the heavy scent from the flowers gave him and his attendants a headache. But on questioning the Kashmiris picking the saffron he was surprised to hear that they had never suffered in this way. He travelled miles to see beauty spots that were recommended to him, and though he was usually delighted—'There were picked fifty kinds of flowers in my presence'—once in a while a place was an anti-climax: 'Although here and there flowers of various colours had bloomed, yet I did not see so many as they had represented to me, and as I had expected.' He was impressed by the huge size of some of the chenar trees, and when he found one that was hollow, amused himself seeing how many people could be squeezed inside.

In fact Jahangir's Kashmir diary is almost a guide-book in itself, the only sadness is that some of the places that he describes are difficult to identify—which is the upland meadow, for instance, that he called 'the place most worth seeing in Kashmir', and which is the pool that inspired him to compose this little poem:

> So clear the water that the grains of sand at the bottom
> Could be counted at midnight by a blind man.

Jahangir had a daughter by a Kashmiri princess, but the baby died when she was only a year old, and the affair does not seem to have been particularly important to him for it rates only a brief mention in his diary. The real love of his life was a Persian beauty known as Nur Mahal or the Light of the Harem. She was later promoted by Jahangir to the name Nur Jahan, which means Light of the World, and was to prove as great a benefactress to Kashmir as any local princess could have been. Jahangir's love for Nur Jahan was so intense and devoted that it turned them into a legendary couple, a sort of Mughal Romeo and Juliet, but much older, for at the time of their marriage he was forty-two and she a widow of thirty-four, shrewd but intelligent and artistic.

Nur Jahan took to Kashmir every bit as enthusiastically as her husband, and during the sixteen years of their marriage the royal couple spent long summers planning all sorts of pleasure grounds and pavilions in the valley. Energetic and creative, Nur Jahan seems likely to have been the brains behind most of these projects. One of these was her own special garden called Darogha or Jarogha Bagh on the banks of Lake Manasbal, near

Safepur. There are only ruins of terraces there now, but it is still a lovely spot. The land juts out into the lake and Constance Villiers Stuart described it as looking like 'some great high-decked galleon floating on the calm clear water.' A few old chenar trees still shade Nur Jahan's garden. The Mughals planted chenar trees in their gardens and to beautify places all over the Kashmir valley, and today the survivors of their trees serve a most useful purpose, acting as flags, pin-pointing the sites the great emperors considered pleasing. Wherever you find a huge old gnarled chenar tree today, you can be sure that there was once a garden there, or a stopping place, or perhaps just a view that the Mughals liked and wanted to improve on.

Sadly, in all Kashmir now there is only one building left to testify to Nur Jahan's refined taste and unerring eye, and that is the mosque she built on the Jhelum river in Srinagar, the Pathar Masjid or Mosque of the Slipper. This simple, elegant building has had a sad history, for it has been used more often as a grain-store than as a place of worship. Tradition has it that this was Nur Jahan's own fault, for when someone asked her how much the mosque had cost to build she simply pointed to her jewelled slippers and smiled and said 'as much as those'—and after that the mosque was deemed unworthy—though happily it is being used as a place of worship today. Architecturally, if not spiritually, the mosque is most successful, and it is one of the three buildings in Kashmir that the Archaeological Survey of India described in their report of 1906–7 as being 'unsurpassed in purity of style and perfection of detail' by any buildings in Agra or Delhi. (The other two are: the lovely grey mosque of Akhund Mullah Shah on Hari Parbat hill that has been mentioned already, and the black marble pavilion in the Shalimar garden.) The Pathar Masjid sits in a rather bleak garden but its grey limestone arches are immaculate, and it is interesting to stroll down to its dilapidated jetty and look at the river view (notice also, the odd-cut stones patching the embankment wall.) Visitors can arrive, like Nur Jahan herself must have done, by boat, but it might be more practical to take a taxi there and combine several sights in one trip. The Shah Hamadan mosque almost faces Pathar

Masjid across the river, the tomb of Zain-ul-Abidin is not far, and neither is the Jama Masjid—but just browsing round this area, crossing and re-crossing the river by the old bridges, makes a fascinating hour or so.

In spite of the Archaeological Survey's remarkable recommendation, most visitors to Kashmir never see the Pathar Masjid, or the Akhund Mullah Shah mosque, but, for reasons quite unconnected with architecture, they go in droves to look at the Shalimar pavilion. The Shalimar garden is the one place in Kashmir that everyone has heard of, and this, whether they know it or not, is all because of a romantic Victorian poem called 'Lalla Rookh', a love story partly set in the Shalimar garden. The author of the poem had never been anywhere near Shalimar, nor indeed Kashmir, in his life, but his poem enshrined them both in the popular imagination forever. And then came another love poem with the memorable line—'pale hands I loved beside the Shalimar'... and so it went on. Shalimar even became the name of an early French perfume.

The famed Shalimar garden is about six kilometres from Srinagar, set between the hills and the smooth waters of the Dal lake. Jahangir and Nur Jahan chose the site because of its natural beauty, and the availability of spring water for irrigation and decoration. 'Shalimar is near the lake', he wrote. 'It has a pleasant stream which comes down from the hills, and flows into the Dal Lake.' He put his son, the future Emperor Shah Jahan, in charge of damming up this stream to make a waterfall, and laying out a garden 'which it would be a pleasure to behold', and the prince was obviously successful because his father described the finished garden as 'one of the sights of Kashmir'

The Shalimar garden was—is—divided into three sections. The lower terraces nearest to the lake and the entrance were public gardens, above came the emperor's own garden which still has the remains of his Turkish bath built into the enclosing side wall, and a Hall of Public Audience in which the emperor's throne is set above a waterfall. There was once a Hall of Private Audience on the terrace above, but the pavilion has disappeared and only its stone foundations and another marble throne are left. Finally, on the highest terraces, far removed from the entrance and guarded by two sentry posts, there was the ladies' garden. This is dominated by the fine black marble pavilion praised so highly by the Archaeological Survey. It was built not by Jahangir, but by Shah Jahan, when he became emperor.

Bernier is the only European to have seen the garden and its pavilions in their prime. He was there only five years after Shah Jahan died, and he wrote that Shalimar was the most beautiful of all the lovely gardens round the lake, with its long fountain-filled canals and cascades and delicate summer-houses set, like islands, in the middle of ornamental pools. He describes the pavilions as having extraordinarily valuable doors—each one made of a single, carved slab of stone. These, he learnt, were taken by Shah Jahan from one of the old Hindu temples in the valley. The doors have long since disappeared, the roofs of the remaining pavilions have been replaced and no one is quite sure what the originals looked like, and the exquisite painting and gilding which once covered the walls has been replaced by modern work of lesser quality. The road that now circles the Dal lake has spoiled the original entrance to the garden. Shalimar was meant to be approached by water, and the long canal that leads to it from the Dal lake was once very much a part of the garden. Now the road has separated the two—the approach canal has rather gone to seed, and visitors who arrived by boat have to disembark at the end of it and plod on foot across the dusty tarmac to the new Shalimar garden entrance.

Like all the Mughal gardens, Shalimar was laid out and planted according to a formula which has as much to do with the Muslim faith as with gardening. A garden had to be divided up into sections so that, like paradise, it would never become tedious or boring. The canals and streams that criss-crossed the gardens and flowed from terrace to terrace represented the waters of life, and even the plants chosen for the gardens were symbolic. The Cypress tree represented death and eternity, fruit trees were the symbol of life and hope, and so on. But the planners were very well aware of human needs too, and in these gardens there was something to please all the senses. The ear was soothed by the pleasant sound of running water from fountains, waterfalls and stone water-chutes that were specially carved to make the streams they carried ripple and splash. The air was perfumed with the scent of flowers (lilac and narcissus in the spring, roses and jasmine in summer), and cherries, apricots, peaches and grapes were planted to provide delicious fruits to eat. Huge chenar trees gave shade from the sun, and the eye—most spoilt of all the senses—could delight in the mass of brilliant flowers that thrive in Kashmir. All the gardens around the Dal were laid out according to this same plan, but each one is very different in character because of the varying steepness of the slopes on which they were built. For instance, Nishat Bagh, which was built by Nur Jahan's brother not far from Shalimar, is on a much more dramatic incline and has, as a result, far more terraces, and steeper ones. The sight and sound of the water tumbling over so many cascades gives Nishat an altogether more vivacious feeling than stately, peaceful Shalimar.

The road around the Dal has completely ruined the entrance to the Nishat garden, but it has not altogether spoiled the charming view of the garden from the lake. To reach Nishat by water, boats have to pass under one of two simple, stone, Mughal bridges set in the causeway that divides the lake. It is only on the other side of the causeway that the garden comes into view across the still water of a lagoon. Not so long ago there was a pavilion at the bottom of the garden which was prettily reflected in the water. It was there when Constance Villiers Stuart wrote her book on the Mughal gardens, and she describes it as a delightful open summer-house with water running through it and down into the lake below, with decorative fountains, and roses climbing up painted wooden pillars, and the air deliciously scented by the lilacs growing on the terrace beneath. 'On a summer day', she writes, 'there are few more attractive rooms than the fountain hall of this Kashmir garden house.' But in the eighty-odd years since she wrote this, heavy hands have been at work. The pavilion has gone and where it stood is a dull, unshaded, stone terrace with a pool and some pieces of broken carved stone arranged around. Water splashes down from the pool to a pretty little terrace below, but then it is forced to disappear under the roadway. The scent in the air these days is more likely to be that of exhaust fumes than lilacs.

The most romantic part of the Nishat garden now is the great terrace at the back which was the ladies' garden. The retaining wall of the terrace is twenty feet high, and an excellent example of Mughal brickwork. Two derelict gazebos—one at either end of the terrace—look out over the lake and must once have been a beautiful sight, and in the grass between the ancient chenar trees are traces of old walls which indicate that there were many other pavilions and buildings up here. There is a secret, mysterious atmosphere on the ladies' terrace at Nishat—one would not be all that surprised to glimpse a bright group of harem beauties, or to hear phantom laughter from a seraglio picnic party under the trees, nor even to catch sight of the stoic figure of the Victorian lady, Honoria Lawrence, sitting on her camp stool writing letters home to her son with an aching heart.

Honoria Lawrence was the first European woman to visit the Kashmir valley. In old books she is always referred to as 'The First White Woman into Kashmir'. She camped in the Nishat garden with her husband, Sir Henry Lawrence, in 1850, and from there she wrote to her much-loved and sorely-missed twelve-year-old son Alick at school in England:

On such a day as this I long for you beyond the power of words to express, that you might share with us the beauty almost beyond that of earth, of the scene we are in. It is noon, but under the broad shade of the Chenar trees we do not feel the sun. We are in one of the old imperial gardens formed in terraces between the mountains that surround part of the lake and the lake itself, a slope of about a quarter of a mile. Down this a stream flows in an artificial channel bordered on each side with gigantic chenars. At each successive terrace, the water tumbles down an artificial cascade. The whole length of the stream is studded with fountains. Above each flight of steps leading from terrace to terrace, something of a pillared portico, with rooms on either hand.... To my right I look up the stream to a line of swelling hills ... to my left I look through the pillars and still follow the stream with its bordering chenars till it is lost in the lake. Next come two tiny islets connected by a bridge, the arch of which terminates the vista.

Honoria Lawrence's description is particularly intriguing for there are no 'pillared porticos, with rooms on either hand' on the terraces at Nishat now, and no sign that such things ever existed save for the marble benches that sit astride the central waterway on nearly every terrace, and are considered a particular feature of this garden. Thanks to Lady Lawrence—and to an earlier traveller, Baron Hügel, who explored Nishat in 1835 and wrote that he particularly liked 'the highly ornamented pavilions built on arches over the canals'—we can assume that these benches must once have been shaded by graceful Mughal gazebos. How rare and magnificent Nishat must have looked then.

Jahangir and Nur Jahan explored the Kashmir valley most thoroughly, and the emperor counselled visitors to follow his example and not to cling exclusively to the town of Srinagar and its environs. In particular he urged them to travel in the countryside upstream of the Jhelum river, towards the town of Anantnag, which he considered more beautiful than that below Srinagar. 'One should stay some days in these regions and go round them so as to enjoy oneself thoroughly', he wrote. Most travellers cannot follow his advice—in a three-day stay, which seems to be what many tour operators allow for Kashmir, they simply do not have time. But Jahangir was right. The country upstream of the Jhelum from Srinagar is charming and much can be seen, even in the course of one day. There is the Jhelum river itself, and pretty, unspoilt villages; in the autumn there are the fields of purple saffron crocuses around Pampur and in spring brilliant yellow mustard everywhere else; there is the tiny, well-preserved temple at Pandrethan and the great temple ruins at Avantipur and Martand, and there are two more Mughal gardens.

The whole of this part of the Kashmir valley is dotted with natural springs, the largest of which were considered sacred by the Hindus of the valley long before the Mughals came—indeed, they still attract pilgrims, particularly on certain auspicious days of the year. The spring at Mattan has already been mentioned, and there is another at Anantnag and another at Kokernag. But the two most dramatic are the springs at Achabal and Verinag, and around these two gushing, bubbling sources of water Nur Jahan and Jahangir were inspired to create beautiful gardens. At Achabal, where the spring comes out of the ground with particular energy, they made an especially watery garden. It begins with a roaring cascade from which the water passes through a network of canals, chutes and wide, still, pools ornamented with rows of fountain jets. In Mughal times, Bernier tells us, this cascade was a marvellous sight at night when it was lit from behind by little oil lamps placed in the wall niches. The main pavilion at Achabal was built across the central stream, and another smaller one was placed in the middle of the first pool—from there the occupants could gaze hypnotically at the tumbling waterfall. The present, rather handsome, buildings are Kashmiri ones (incorporating some fine pieces of Mughal stonework) on Mughal foundations. A few gnarled old chenars are all that is left in a garden that once burgeoned with flowers and fruits and trees of all sorts, but it is still a pleasant place. To the left of the garden as you walk down, there are the interesting remains of a Mughal hammam or bath house. The baths and their surrounding stone slabs are almost intact and it is possible to identify the hot room by the double floor which must have contained the fire for under-floor heating.

But it was at Verinag, a spring towards the end of the Kashmir valley not far from the Banihal pass, that the royal couple designed the palace and garden which they came to love more than anywhere else in Kashmir. Pine-covered hills rise sharply behind the spring which surges up into a deep pool of emerald water, and it is not hard to see why Verinag was always considered a sacred place; nor why Jahangir fell in love with it when he visited it, twice, with his father Akbar; nor why he decided to build a palace and garden there. 'When I was a prince', he wrote afterwards, 'I had given an order that they should erect a building at this spring suitable to the place. It was now completed. There was a reservoir of an octagonal shape . . . round it halls with domes had been erected and there was a garden in front of them . . . round the reservoir was a stone walk . . .'. Later on he added that his builders had completed another garden with a canal at Verinag, and had built halls and houses there, and that 'in the whole of Kashmir there is no sight of such beauty'. Today only the octagonal reservoir with its stone walkway and arched recesses is left of what was the fountain court of Jahangir's palace. Inside some of the recesses are small flights of steps which must once have led to rooms above, and there are the remnants of some finely carved stone brackets jutting from the wall which may have supported balconies. Two original stones, inscribed in Persian, are set in to the wall along the walkway—one gives the date of the building, 1609, and the other tells the name of the builder, 'constructed by Haidar, by order of Jahangir'.

According to Constance Villiers Stuart, who saw Verinag early this century when more of the buildings were still standing, this fountain court lay behind the main part of the palace which was built above the arch through which the water flows out of the reservoir, and extended along the top terrace of the garden. (This is where most of the rubble is lying today.) The watercourse which comes in from the side of the garden to join the main canal, once ran directly in front of this palace.

It was logical that having transformed the Kashmir valley into a delightful retreat with summer houses and gardens in all their favourite places, Jahangir should turn his mind to making the long journey to Kashmir from India more comfortable.

In Mughal times there were several routes into Kashmir. The most northerly one followed the course of the Jhelum river, entering Kashmir via Baramula, from where the royal party could travel on to Srinagar either by boat or overland. A southern route came up via Jammu, over the Banihal pass and on through the length of the Kashmir valley to Srinagar. And a middle route began at Bimber, wound over the Pir Panjal mountains and into the valley. At first the northern route was the most popular with the Mughals, who built a beautiful garden—a kind of base camp for expeditions into Kashmir—at Wah Bagh near Rawalpindi. Later the middle route over the Pir Panjal mountains became the favourite and it was along this, the Royal Route or Imperial Route as it became known, that Jahangir decided to build a chain of caravanserais to shelter the royal travellers: 'I had directed that from Kashmir to the end of the hilly country buildings should be erected at each stage for the accommodation of myself and the ladies, for in the cold weather one should not be in tents.' The first time the new buildings were used there was a great disappointment: 'Although the buildings at this stage had been completed', writes Jahangir, 'as they were still damp and there was a smell of lime, we put up in tents.' Curiously enough, exactly the same sort of thing happened to another royal traveller in the area about 250 years later. When the Prince of Wales, the future King Edward VII, visited Jammu in 1876 a palace was erected for him to stay in. It took three months to build and cost a great deal of money, but after only two nights in it he moved out because of the damp and slept in a tent.

Honoria Lawrence travelled along this Royal Route into Kashmir in 1850 and as she was bumped along in her *dhoolie*—a sort of rough version of a Sedan chair—she thought regretfully of the 'palmy days' of the Mughal emperors who had often made this very same journey, and wrote to her son in England:

> of all the pomp of that time nothing remains, save the halting places built at about every ten miles for the Emperor. These exist still, some in absolute ruins, some partly habitable, all built on the same plan: a large open quadrangle, corridors, serving for stables and such like running along two sides, the side facing the gateway containing the royal apartments. The material is stone with a small proportion of brick.... Ghulab Singh has this year built places for European travellers. Oh what a falling off! In one corner of the old Serai enclosure are four earthen walls with earthen floor, rough plank ceilings and doors; such is the place that now felt a welcome shelter.

Frederic Drew followed in the footsteps of the Mughals and Honoria Lawrence when he travelled the Royal Route in 1862 to take up a post in Kashmir, and he paid particular attention to Jahangir's serais along the way. The first one, at Bimber, was the starting point for the Mughal court's journey over the mountains, and the serai had to be a large one to accommodate the vast assembly. Drew measured that it was 300 feet square, and strongly built of stone and brick. Further along the route, at Saidabad, Drew came upon what he considered the finest of all the serais—'massively built', with three courtyards, baths, a terrace and many arched and vaulted rooms. At Naoshera there was a well-preserved serai with an inner courtyard; at Changas another magnificent stone-built serai; at Rajaori more Mughal relics; at Thanna the remains of a marvellous courtyard. In fact, as Drew pointed out, 'at most or all of the stages out to Kashmir are some remains of the royal rest-places.'

When the Jhelum Valley Road—the first road into Kashmir that could take wheeled vehicles—was completed in 1890, the Royal Route over the mountains ceased to have any importance at all and photographs taken of the Mughal serais fifty years later, in the 1940s, show them in a sorry state of dereliction. Today it is impossible for travellers to use the old road as Bimber lies only a few miles from the Pakistan border, and the early part of the route goes through restricted military areas.

In 1627 Jahangir and Nur Jahan spent what was to be their last summer in Kashmir. They should not have gone there at all, for Jahangir's son, the future Shah Jahan, was in open rebellion against his father and the emperor's throne was not entirely secure. However, Jahangir was unwell. His health had been failing for some time, and the royal couple must have believed that a holiday in Kashmir's cooler climate would make him better. Jahangir survived the summer in the valley but the court had not travelled much of the long road back to Lahore when he became very ill and died at a village called Baramgala in the mountains. This was perhaps appropriate, for Baramgala was a place that Jahangir had particularly loved. There was a waterfall there and he had, on a previous journey, ordered that a terrace should be built on a rock overlooking it, and a stone put up to commemorate the date of that visit.

It is said that as Jahangir lay on his deathbed his anxious courtiers asked him if there was anything that he wanted. 'Only

Kashmir', he replied in a whisper. But even Mughal emperors' wishes were not always granted, and it was thought more politic to carry his body back to India. So that it would last the course, his entrails were removed and buried in the courtyard of the Changas serai. His body was eventually laid to rest in Lahore where Nur Jahan spent some of the eighteen remaining years of her life planning his tomb. His son, the rebellious prince, who in happier times had laid out the Shalimar garden according to his father's instructions, now became Emperor Shah Jahan.

Shah Jahan did not share his father's passion for Kashmir, possibly because he was more Indian than Jahangir (both his mother and his grandmother had been princesses from Rajasthan). He did not have that fierce nostalgia for the mountains that drove Jahangir to Kashmir so often. Shah Jahan was content to build his architectural masterpieces, the Taj Mahal and the Red Forts of Delhi and Agra, against the backdrop of dusty plains. Nonetheless, though he visited Kashmir only four times after his accession, Shah Jahan did not neglect the valley and was responsible for one of the prettiest of the gardens around the Dal lake, the Chashma Shahi. This is a small garden compared to Nishat and Shalimar, with only three terraces, but it is built high on a steep slope and has wonderful views of the lake and the hills. Chashma Shahi means Royal Spring, and the garden was planned around a natural spring which gushes out of the hillside at the top of the garden and which is still drunk reverently by visitors who consider it holy. In Constance Villiers Stuart's day the spring bubbled up into an exquisitely carved marble basin, but this has vanished and an ugly concrete and glass contraption covers it now—though there is a prettily carved marble basin further along the stream. Mrs Villiers Stuart also describes a pretty open pavilion at Chashma Shahi with a stone floor under which water flowed to cool the inhabitants before dropping down a carved stone chute twenty feet to the terrace below. There is no pavilion there today, sadly, but its stone floor remains as an open platform, and the water still runs underneath it and is carried down to the next terrace by the pretty carved chute.

Shah Jahan appointed his own son, Murad, as Governor of Kashmir in 1640, and during this prince's year of office he not only married a Kashmiri girl but is also credited with building a pretty pavilion in the shade of four chenar trees on the tiny island of Son Lank in the Dal lake near Hazratbal mosque. This island, which is often called the Char Chenar or Four Chenars, and its charming Mughal pavilion, are mentioned by nearly all the early European travellers to Kashmir. Indeed, some of them camped in the slowly-deteriorating building, and it can clearly be seen in a sketch done in 1835 by Godfrey Vigne, a British traveller. Today the pavilion has gone, there are only two large old chenar trees left, and two small young ones, some picnic shacks and a public lavatory—but there are traces of handsome old stone landing stages on each side of the island.

Across the lake, below Chashma Shahi is another island called the Rup Lank, which is also, confusingly, sometimes known as the Char Chenar. Both Zain-ul-Abidin and Jahangir are supposed to have built summer-houses on it, but not a trace remains of either.

Luckily for today's visitors, a garden and a mosque that Shah Jahan's eldest son, Dara Shikoh, created in Kashmir do still exist, for they are among the loveliest of all the Mughal relics there. The garden is a magical, mysterious place, perched on the shoulder of a hill far above and aloof from all the other pleasure gardens around the Dal lake, and it fully deserves its romantic name of Pari Mahal or Fairy Palace. Pari Mahal was not intended to be a mere pleasure garden—it was built as a school for Dara Shikoh's spiritual guide and counsellor, Akhund Mullah Shah, and perhaps that explains its pleasing atmosphere.

Even now, the remains of the eight steep terraces, with their fine retaining walls built in stone with decorative arches, are most impressive. No wonder some of the European visitors in the nineteenth century chose to pitch camp here. A square building on the top terrace dominates the garden. Its roof has been levelled off and flattened, but it is believed that it was originally domed, and might have been some kind of observatory. Two terraces below, there is the entrance to the garden and many of the original buildings in surprisingly good condition. Two terraces below that is a charming dove cote and two octagonal gazebos that house the stairs to the next level. Traces of canals and pools and pipes can be seen everywhere, and spring water must once have coursed delightfully down the garden. Until quite lately, Pari Mahal was abandoned and overgrown and only snakes went there, but the terraces have now been tidied up, ugly plastic lamps have been installed all over the place, and an ornamental tank has been 'reconstructed' on the second terrace which seems to be of all the wrong proportions. Pari Mahal was far more romantic in its overgrown days, but nothing could really spoil it, the views are too breathtaking, the atmosphere too pleasing. Sensitively and accurately restored, and replanted with imagination Pari Mahal could become one of the sights of India.

Very different to Pari Mahal, but equally marvellous in its own way, is the mosque that Dara Shikoh built for the same Akhund Mullah Shah on the slopes of Hari Parbat hill (this is the mosque already mentioned in the Introduction and earlier in this chapter). It stands inconspicuously to the left of the walkway leading up to the *ziarat* of Makdhum Sahib—the tomb of a Muslim holy man and a tremendously popular shrine, visited especially by people hoping for a cure for an illness. The Akhund Mulla Shah mosque is a small, compact grey stone building in the same simple elegant style as Nur Jahan's 'Slipper Mosque', and every part of it is refined and polished and finished to perfection. The only tragedy is that apart from the writers of the Archaeological Survey of India in 1906–7, no one seems to have appreciated this building and it has been neglected and vandalized to the point where, if something is not done about it soon, it will be too late. The notice-board in the grounds which describes the mosque says 'the stone finial above the pulpit is the only remaining one in Kashmir . . .', but already the finial has gone.

Dara Shikoh's third creation in Kashmir was a garden at Bijbihar on the Jhelum river, not far from the town of Anantnag. His design was unlike all the other Mughal gardens of Kashmir for he built it on opposite banks of the river so that the terraces faced each other across the water, and a stone bridge connected the two halves. The bridge has long since fallen down, and the pavilions have crumbled, but the watercourses he planned, and some of the great chenar trees he planted are still there to bear witness to this most original garden.

Dara Shikoh was a gentle, studious man who, like his grandfather Jahangir, loved nature and art. He clearly shared Jahangir's love for Kashmir too, and it is tantalizing to speculate on what he might have achieved in the valley as its emperor, when, as a mere prince, he created three such beautiful places as Pari Mahal, Akhund Mullah Shah mosque, and Bijbihar garden. But in the struggle to succeed Shah Jahan, Dara Shikoh, though the eldest son and the rightful heir, lost the throne—and his life—to Aurangzeb, his very orthodox and puritanical brother.

The emperor Aurangzeb visited Kashmir only once, in 1665, and left nothing there to be remembered by except rules—rules which harassed the Hindu population, and rules which forbade music and dancing and drinking which were particularly painful to the exuberant Kashmiris.

For most of his reign Aurangzeb was preoccupied with problems far away from Kashmir, at the other end of his kingdom in the Deccan, where he spent the last years of his life campaigning ceaselessly against turbulent local chiefs, trying to keep his empire intact. He succeeded, more or less, but his heirs were weak and quarrelling men, and after Aurangzeb's death in 1707 most of the power of the Mughal emperors was simply frittered away.

In Kashmir a golden age had come to an end. No longer was the valley the jewel in the Mughal crown, the cherished retreat of the great emperors to be favoured and protected above all other parts of their empire. After Aurangzeb's single visit the Mughal court never made the long journey there again, and the governors appointed to rule Kashmir could now rarely be bothered to go to the valley themselves, but appointed representatives to rule in their places—something that would have been unthinkable under Akbar, Jahangir or Shah Jahan. Once again Kashmir became isolated from the rest of the world and at the mercy of greedy local officials who could behave as they pleased, and the all-too-familiar story of corruption, oppression and civil strife began all over again. It was at around this time that the Nehru family, forebears of three Indian prime ministers, emigrated from Kashmir to Uttar Pradesh.

In 1739 there was terrible public proof of the weakness of Mughal rule in India, when the Persian leader Nadir Shah marched an army all the way to Delhi, sacked the city, massacred more than 30,000 people, and carried away to Persia a vast hoard of treasure including the Peacock Throne and the Kohinoor diamond. Nadir Shah did not even bother to kill the humiliated emperor, and greatly weakened, emperors remained on the throne in Delhi until just after the 1857 'mutiny'. Nadir Shah's brutal but successful raid acted like a signal for all the little states that had been welded together into an empire by the might of the Mughals, to seek their independence. The Mughal empire began breaking up, and, waiting on the sidelines were plenty of people longing to pick at the pieces. The well-watered, fertile plains of the Punjab—the part of India nearest to Kashmir—were an especially desirable portion, and the Punjab became a battleground between the Sikhs, who had newly become a force to be reckoned with, and the Afghans, who had acquired a dynamic new leader, Ahmad Shah Durrani. In 1752 the Afghans swept down from their mountains into India in their third attempt to wrest the Punjab from the Sikhs, and this time they succeeded beyond their hopes by capturing not just the Punjab, but the valley of Kashmir as well.

Fortunately, though there were so few early European travellers to Kashmir, their visits conveniently coincided with the three main periods of foreign rule in the valley, and so we have eye-witness accounts of Kashmir under the Mughals, the Afghans, and the Sikhs. George Forster, an Englishman, travelled to

Kashmir in 1783 and described the sorry state of the valley under Afghan rule—the cruelty, the extortion and persecution. He is quoted in the next chapter, so there is no need to dwell on the gory details here except to give one neat illustration of Afghan greed. Forster wrote that in the reign of Aurangzeb the taxes collected in Kashmir amounted to Rs 350,000, but that at the time of his visit 'not less than twenty lakhs [Rs 2,000,000] are extracted by the Afghan governors.'

Only two Afghan governors actually seem to have contributed anything to the valley. One was Amir Jawan Sher, who was in power in the 1770s. He built the Amira Kadal bridge in Srinagar and the Shergarhi palace—which was later rebuilt and is now a government building known as the Old Secretariat. He laid out some gardens too, but is also supposed to have ordered the destruction of some of the Mughal buildings, including Akbar's palace on Hari Parbat hill. On Hari Parbat today there is only an overgrown and neglected spot called the Darshani Bagh or Garden of Audience, which is said to be the remains of Akbar's palace garden.

The only really respectable Afghan governor was the last one, Atta Mohammed Khan, whose administration from 1806 to 1812 has been described as an oasis in the desert of Afghan rule. Atta Mohammed Khan built the fort that stands today on the summit of Hari Parbat hill, where his name can still be found, inscribed over the fifth gate.

In the mean time, back in Afghanistan itself there had been violent power struggles between rival factions of the ruling family ever since Ahmad Shah Durrani had died in 1773, and eventually these led to their losing Kashmir. For in 1818, the Afghan governor of Kashmir, hearing that his brother had been murdered by the king in Kabul, rushed home to avenge his death, taking with him the best part of the army; and the Sikhs, seeing the valley they had coveted for years now almost defenceless, attacked. The Afghans were driven out, the Sikhs took over, and Kashmir found itself in 1819 out of the frying pan and in the fire.

Sikh rule in Kashmir lasted only twenty-seven years, but they were terrible ones for the country. The new men looked down on the Muslim Kashmiris with contempt, and treated them literally worse than cattle—for you could kill a Kashmiri and get away with a fine, but now the cow became sacred and the killing of one punishable by death (a sentence which was carried out quite often for the people were desperately poor and hungry—a family of 17 souls was once burnt alive for killing a cow). The mosques were closed, tax collectors took nine-tenths of the crops produced, and then the government sold the produce back to the people at extortionate rates.

As if Kashmiris did not already have enough to bear, nature chose to be particularly unkind to their area during the years of Sikh government. There was a severe earthquake, a cholera epidemic and, most terrible of all—between 1832 and 1834—an appalling famine which decimated the valley and, it is said, reduced the population from 800,000 people to only 200,000. This seems to have been unacceptable, even to the Sikhs, and a new governor, Colonel Mehan Singh, was sent up to sort out the situation. Colonel Mehan Singh was not a bad governor—he organized famine relief, tried to revive the trade and industry of the valley, and even built a garden, the Besant Bagh, near the Shergarhi palace—but he did not greatly impress the foreign visitors who met him. Godfrey Vigne, an English traveller, wrote: 'He was the fattest man I saw in the East, with a good-humoured aspect, and the air of a bon vivant. How he contrived to exist in good health I knew not. At breakfast he ate largely of almonds stewed in butter; and never went to bed sober by any chance.'

Baron von Schonberg, a visitor to Kashmir towards the end of the Sikh period, wrote: 'I have been in many lands, but nowhere did the condition of the human being present a more saddening spectacle than in Kashmir.' Godfrey Vigne was equally horrified at the suffering of the Kashmiris, but he confidently thought that it was only a matter of time before the British, who by now had the upper hand in India, marched in, planted their flag 'on the ramparts of the Hari Parbat' and developed the valley's many resources so that it became the 'sine qua non of the oriental traveller . . .'.

The leader of the Sikhs, Ranjit Singh, 'The Lion of the Punjab' as he was called, never visited the valley himself, and, unlike the Mughals, his reasons for wanting to possess it were purely connected with what he could squeeze out of it in terms of revenue and treasures. He owned a vast collection of Kashmiri shawls, and seems to have found Kashmiri women attractive (and vice versa), for an English officer who attended his cremation wrote that though 'everything was done to prevent it' his four wives flung themselves on the funeral pyre and were closely followed by 'five of his Cachmerian slave girls'.

Ranjit Singh was something of a genius. At the age of seventeen, although he was illiterate and blind in one eye from smallpox, he had become chief of a small Sikh clan in the Punjab, and had gone on to unify all the hostile, warring Sikh groups into one powerful nation. He had pushed the Afghans out of most of the Punjab, acquired the Kohinoor diamond from them while doing so, and had added Kashmir and other small mountain kingdoms to his dominions. But when he died in 1839 the Sikhs began to fight among themselves again, and two of Ranjit's sons and two grandsons were assassinated in the struggle for power. In the end his youngest child, Duleep Singh, was placed on the throne, though he was only six years old.

Ranjit Singh had always been shrewd enough not to provoke the British, not to prod or needle them for fear that they would turn on him. In fact he had parleyed with successive British governor-generals, meetings at which both sides tried to outdo each other in splendour and magnificence. Victor Jacquemont, a French traveller in India at the time of one of these encounters, wrote that the press reports on it were disgracefully hypocritical: 'Not one of these papers, for instance, has dared to say that, on his second visit to Lord William Bentinck, Runjeet gravely committed a nuisance in the corner of the superb tent. . . .'

After Ranjit Singh's death, however, the powerful Sikh army became more and more restless and troublesome and worrying to Duleep Singh's mother and the ministers who ruled in his name, and they thought it expedient to encourage the army to attack the British, for at least that would keep it from meddling in politics in Lahore. The battle that followed became known as the First Sikh War, and it was only narrowly won by the British.

The British did not take over the Punjab after their victory—that happened after the Second Sikh War which followed within a couple of years. But Sir Henry Lawrence was appointed British Resident in Lahore, and a heavy fine was imposed on the Sikhs. Unable to pay this fine in cash, the Sikhs offered instead to give Britain the little mountain states of Kashmir, Jammu, Ladakh and Baltistan, which were part of Sikh territory. Kashmir apart, these states were administered by a local raja, Gulab Singh, who had been the great favourite of Ranjit Singh. Gulab Singh had no intention of losing his lands to the British if he could help it, so he quickly offered to pay the fine on behalf of the Sikh nation—on condition that he was allowed to keep the states he administered forever, and add Kashmir to them.

The British agreed, and on 16 March 1846 a treaty was signed in which

> The British Government transfers and makes over, for ever, in independent possession, to Maharaja Gulab Singh, and the heirs male of his body all the hilly or mountainous country, with its dependencies, situated to the eastward of the river Indus and westward of the river Ravi....

In exchange, Gulab Singh agreed to pay the British Rs 7,500,000 straight away, plus, as an annual token acknowledgement of British supremacy, 'one horse, twelve perfect shawl-goats of approved breed (six male and six female) and three pairs of Kashmir shawls.' Gulab also pledged the loyalty and support of his armies to the British should they ever need them.

Now Godfrey Vigne's vision of the British flag fluttering over a Kashmir of peace and prosperity could never be realized and many people were bewildered—as Vigne himself must have been—at Britain's sale of Kashmir to Gulab Singh. The explanation for it seems to have been that the deal provided Britain with some urgently needed cash, as well as guaranteeing the security of parts of India's northern border. These were rugged, remote areas that would have been difficult for the British to defend at the time, so, they must have reasoned, why not let Gulab Singh do it for them.

Baron von Schonberg was astonished that the British could put their trust in Gulab Singh:

The natives in the northern province maintain that the English are not acquainted with his character, or that they would not for one day keep up a friendly intercourse with him.... How often have my people, around the crackling fire, when assured that the ear of no spy was listening, for hours together, told us of this tyrant, this betrayer, to whom no duty, no promise, no sorrow was sacred—who cared for no human being.... He would sell the Punjab ten times over to the English to serve his own purpose and would betray the English then, did he find a profit in so doing.

But it seems that the British had no illusions about what they were doing. Sir Henry Lawrence, who had been one of the signatories to the treaty, wrote: 'I have no doubt that Maharaja Gulab Singh is a man of indifferent character; but if we look for perfection from native chiefs, we shall look in vain ... he has many virtues that few of them possess, viz. courage, energy, and personal purity ...'.

Gulab Singh's family were Dogras—which is the name of the people who come from the foothills of the Himalayas between the rivers Chenab and Sutlej—and they were Rajputs, which means they were of the Hindu warrior caste. The Dogras have always had a fierce fighting reputation—indeed the very first Victoria Cross awarded to the Indian army went to a Dogra sepoy for bravery in Mesopotamia. Gulab's people were not well off, so he and his younger brothers had enlisted as soldiers in the service of Ranjit Singh. In an extraordinarily short time the three brothers' military prowess and good looks attracted the attention of Ranjit Singh himself, and they were promoted to high positions, close to the Sikh leader. Eventually all three brothers were awarded the title 'Raja' and granted lands, though, of course, they still owed allegiance to Ranjit Singh. Gulab Singh was given Jammu, one of his brothers was given Punch nearby, and the other a place called Rajaori in the same area. Before Ranjit Singh's death Gulab Singh had invaded Ladakh as well, and shortly afterwards he captured Baltistan to the north of Kashmir. Now the three brothers held a solid block of territory in the hills. Only Kashmir was obviously missing, like the gap left by a lost jigsaw piece, and Gulab Singh began intriguing to gain possession of it as well: It had to be by intrigue, for not even Gulab Singh would have dared to try and seize the valley by force from his old patrons and allies, the Sikhs.

Realizing that it was the British he needed on his side, Gulab Singh managed to keep himself out of the Sikh War. Instead he became an intermediary between the two sides, and later an advisor to Sir Henry Lawrence—and therefore in a perfect position to suggest the compromise over the fine imposed on the Sikhs.

Once the treaty with the British was signed, Gulab Singh marched to Kashmir to claim the country he had so cleverly manoeuvred into his hands, but to everyone's surprise there was fierce resistance from the Kashmiris who drove out Gulab Singh's Dogra troops. Gulab Singh was forced to appeal to his new friends, the British, for help, and Sir Henry Lawrence plus a large force of Sikh soldiers went to Kashmir to assist with the takeover. Gulab Singh, the first Maharaja of Kashmir, was at last able to enter Srinagar on 19 November 1846. His son, his grandson, and his great-grandson succeeded him, and the Dogra dynasty, as it is called, was in power for 106 years. The last heir to the throne was born in Cannes in the south of France while his parents, Their Highnesses the Maharaja and Maharani, were on an European tour. How far removed from the medieval world of their ancestor, the rough soldier-of-fortune who, by courage and shrewd manoeuvring, had gained for himself one of the loveliest kingdoms on earth.

Gulab Singh's friendship with the British, plus the fact that it became considerably safer to travel to the valley after Britain annexed the Punjab, meant that there was soon a regular flow of western visitors to Kashmir. All of them seem to have found themselves torn between delight at the beauty of the valley and horror at the abject state of the people and the harshness of their ruler. Andrew Adams, a surgeon of the 22nd Regiment and amateur naturalist, spent his leave in Kashmir in 1867, and his description of the approach to Srinagar is typical:

> As the small gondolas glided slowly towards the entrance to this little Venice of Asia our attention was directed to two human skeletons suspended in cages on the river banks; these we were informed, were criminals that had been executed some years before and were left on these gibbets as a warning to all malefactors. We were not altogether unprepared for such examples of Gulab Singh's mode of rule, having read of his horrible deeds. . . .

Under Gulab Singh the Kashmiris were not persecuted in quite the ruthless, random way they had been under the Sikhs, but nonetheless the new ruler was greedy for revenue, the taxes were burdensome, and the lives of the people did not improve. Aggravating the situation was the fact that though the vast majority of Kashmiris were Muslims they were treated as second-class citizens by the maharaja and his officials who were fiercely Hindu. Hindus were promoted and favoured above the Muslims in every way—they were usually exempt from the system of *begar* or forced porterage—and Hindu tax-collectors were often blatantly corrupt. Cecil Tyndale-Biscoe, who came to the valley during the rule of Gulab Singh's grandson, wrote: 'There was such an army of Hindu officials whose duty it was to collect the grain, that, when all had been supplied, both lawfully and unlawfully, there was very little left for the zaminder and his family who had farmed the land.'

However, the new maharaja did at least spend time in Kashmir himself, dividing the year—as does the present government of Jammu and Kashmir—between the two states. The maharaja arrived each year in the spring, stayed in the valley through the summer months, and left for Jammu again in the autumn. Soon these royal arrivals and departures became events of great ceremony and splendour. The houses lining the Jhelum river would be jammed with spectators, the city would be decorated in its traditional way—with shawls hung from the windows, and, if it was spring, the earth roofs of the houses would be bright with irises and tulips in flower.

It was customary for all the state officials—including, later, the British Resident and his staff, as well as senior Europeans living in Kashmir—to leave the city in boats and go to meet the maharaja's little fleet as it made its way along the river to Srinagar. 'The Maharaja arrived this year', wrote Sir Francis Younghusband who was British Resident in 1908,

> on the most perfect day in Spring. Before the time of his arrival the river was alive with craft of every description, from the Resident's state barge of enormous length, and manned by about 50 rowers dressed in scarlet, to light shikaras, and even two motor-boats. As we emerged from the town the banks on either side were covered with fresh green grass, the poplars and some magnificent Chenar trees overhanging the river were in their freshest foliage. And coming up a long reach of the broad glistening river was the Maharaja's flotilla, with their long lines of red and of blue oarsmen giving colour to the scene.

But that was later. In Gulab Singh's day there was no British Resident in Kashmir, nor indeed any other official British advisor—but Gulab appointed a Westerner as Commander-in-Chief of his army, an old comrade-at-arms, Colonel Alexander Gardner. Variously described in his own time as 'a plausible and ingenious scamp' and as 'one of the most extraordinary men in India', Colonel Gardner was a unique character—a sort of real-life version of Kipling's Man Who Would Be King, a scarred and seasoned veteran of many far-flung campaigns. Because of an old wound in the neck, he had to hold his throat with an iron pincer whenever he drank. He was born in America in 1785 of a Scottish father and a Spanish mother, and educated by the Jesuits in Mexico. After various unsatisfactory wanderings around Europe and Russia, Gardner decided to offer his services as a mercenary to the Persian court. Accordingly, he set out for Persia but, being the man he was, he became caught up by events on the journey and, instead, spent the next eight and a half years adventuring in Central Asia, hiring himself out to various local chieftans. The most thrilling adventure story would seem positively insipid compared to his life—he even married a beautiful Afghan girl and had a son by her, and adored them both until they were killed in an attack by enemies of his chief. In fact, by 1832, things were not going too well for Gardener, and he decided to enter the service of Ranjit Singh.

At that time Ranjit Singh employed no less than thirty-nine Western officers and advisors in his army—mercenaries from France, Italy, England, America, Russia, Greece, Spain and Germany—as well as three doctors from Austria, England and France. It is, perhaps, no wonder that Ranjit Singh's forces gave the British such a hard time in the First Sikh War, for at least one of his French advisors, General Allard, and one Italian, General Ventura, had been officers in Napoleon's army and fought at Waterloo. His European officers had to obey certain Sikh tenets—they were not allowed to eat beef or shave their beards, but they were given a dispensation to smoke, and the more senior of them lived in fantastic style. General Ventura, for instance, is reported to have had 'forty or fifty female slaves', though in the end he married a European lady and retired to Paris.

Gardner worked for the Sikhs until the outbreak of the First Sikh War—he would, he wrote, have happily gone into battle against the British—but he was dismissed from the Sikh army by enemies he had made after Ranjit Singh's death. Instead, he cheerfully went to work for Gulab Singh, the new Maharaja of Kashmir, and an old friend from the days when they had both admired and fought for Ranjit Singh. Gardner stayed in Jammu and Kashmir for the rest of his life. He was well rewarded by Gulab Singh and, according to a contemporary, 'he lived in good style, after the native fashion, being from long habit a complete Oriental.' In his old age Gardner had a daughter whom he named Helena—history does not reveal who her mother was. After her father's death in 1877, Helena, now a Mrs Botha, returned to Kashmir for a visit, and old Sikh and Dogra soldiers flocked to pay homage to 'Gordana Sahib's' child, and tell her how much they had loved and revered her father.

Old Gardner, in his extraordinary green and yellow tartan uniform, designed by himself, had become one of the sights of Kashmir before he died. A Captain Seagrave wrote this description of him:

He walked into Cooper's reception-room one morning, a most peculiar and striking appearance, clothed from head to foot in the 79th tartan, but fashioned by a native tailor. Even his pagri was of tartan, and it was adorned with the egret's plume, only allowed to persons of high rank. I imagine he lived entirely in native fashion: he was said to be wealthy, and the owner of many villages.

Gulab Singh died in August 1857 and, according to Colonel Gardner, the old maharaja's last words to Ranbir, his son and heir, were, 'Should one Englishman be left in the world, trust in him.' In fact, it was the British who were in sore need of someone to trust at that moment. The Indian mutiny had broken out the previous April and in August, when Gulab Singh died, they were trying, unsuccessfully, to recapture Delhi from the mutineers. Almost Ranbir Singh's first act on becoming maharaja was to send a force of Dogra troops with 200 cavalry and six guns to help the British forces. After Delhi was recaptured and the mutiny suppressed, the British knighted Ranbir Singh in gratitude, and altered the treaty between his family and the British government so that, in the event of there being no direct male heir, Kashmir could pass to a nephew or cousin. They also added two guns to his ceremonial salute, making it twenty-one.

Ranbir Singh was extremely hospitable to European visitors. Every new arrival was called upon by a representative of the maharaja with assurances of welcome and presents of food and fruit and a sheep or a goat. Dr Wakefield, who visited Kashmir in 1875, wrote:

> The politeness of the ruler of the state to strangers is frequently extended not only to the living but to the dead; for if an English officer is so unfortunate as to come to his death during his stay in the Valley, this attentive prince, to show his sympathy with the melancholy event, usually sends a shawl of price to wrap the body in before burial.

When a maharaja died his body could be wrapped in as many as forty layers of shawls.

But for all his hospitality, Ranbir Singh was fiercely independent, and would not tolerate any interference in the way he ruled his state. He refused to allow the British to install a Resident (their usual way of keeping a hand in the affairs of the princely states) and, though he built houses for them to lodge in, he would not allow any foreigner to buy land in Kashmir—which is why the famous houseboats came into being.

Ranbir Singh may have been generous to his European visitors, but his own Muslim subjects struggled on in abject poverty, and in the 1860s a visitor called Robert Thorp incurred the full fury of the maharaja's rage for trying to publicize the plight of the Kashmiri poor. Robert Thorp seemed the very model of an English gentleman, but in fact he was the son of a Kashmiri woman who had married an Englishman, and he had been born in Srinagar. It was on a holiday visit to inspect his birthplace that the young man became fired to fight for what were, after all, his own people. Then and there he embarked on a campaign to expose their miserable condition by sending dozens of articles to newspapers outside Kashmir. In these he was quite as critical of the British government as of the Maharaja of Kashmir: 'For purposes entirely selfish, we deliberately sold millions of human beings into the absolute power of one of the meanest, most avaricious, cruel and unprincipled of men that ever sat upon a throne.' Thorp was referring to Gulab Singh, but Ranbir Singh proved himself his father's son by having Robert Thorp murdered when he refused to obey an order to leave the valley. He was buried in the European cemetery in Srinagar, and Cecil Tyndale-Biscoe, a missionary, said that he never passed the young man's grave without raising his hat in respect 'for the mortal remains of Lieut. Robert Thorp who gave his life for the Kashmiris . . .'.

There is a saying that God creates droughts and floods, but that man creates famine, and in 1877 a terrible example of this took place in Kashmir. At that time the annual harvest in the valley was never allowed to begin until the government gave its permission—this was so that all the previous year's rice stocks could be sold off at an inflated price. In the autumn of 1877 the new rice crop was a promising one, the farmers were ready to start bringing it in, but greedy officials postponed the harvest and, before the rice was cut, disaster struck—it began to rain. The rain poured down for weeks, the entire crop was ruined, there was nothing to eat, and for two years Kashmir became a nightmare country of death and disease. So many people starved or ran away that villages were wiped out, whole valleys became deserted, and the population of Srinagar was halved.

In 1885 Ranbir Singh died, his son Pratab succeeded him, and the British seized the chance to install their first Resident in Srinagar. In Pratab Singh's reign, at long last, some steps towards improving the lives of the Kashmiris were taken. One was that a proper road linking the valley with India and the outside world was opened in 1890—the Jhelum Valley Cart Road it was called, and it was built by an intrepid British contractor called Spedding with a gang of labourers of whom Tyndale-Biscoe wrote: 'It would have been difficult to collect a finer lot of scoundrels.' Another was the reorganization of the whole system of taxation in the valley, which in turn meant fixing boundaries and sorting out the rights and duties of tenants and landlords. An Englishman called Walter Lawrence tackled this complicated can of worms, and at the same time he persuaded the maharaja to abolish the cruel custom of *begar* or forced porterage. Lawrence was not surprised to find the Kashmiris sullen, desperate and suspicious, for, he wrote, 'a people so broken and degraded . . . do not in a few years harden into a resolute and self-respecting community.'

He viewed the future with optimism, however, and was particularly pleased to see the increasing numbers of tourists taking holidays in the valley each year: 'The annual inroads of foreign troops who pillaged the country . . . have given place to the welcome invasion of European visitors who spend large sums of money in the Happy Valley . . .'. Indeed, by the turn of the century, Kashmir was a well-established holiday resort for the British in India, and even Kashmiris who had fled the valley in bad times began to drift back to their villages.

Pratab Singh had no heir, and when he died in 1925 his nephew Hari Singh, son of Pratab's powerful younger brother, Sir Amar Singh, became maharaja. Young and handsome, Hari

Singh was very different from the elderly, old-fashioned and devoutly Hindu Pratab. 'He is a clean-cut, highly educated, admirably informed man, and all that shines out of his black, closely set eyes. . . . It is a face you would notice in a crowd, and the figure, dressed in dark brown with tight-fitting Jodhpur breeches and brown shoes, has an equal air of distinction. . . .'—wrote an admiring British visitor who met Hari Singh at a tea-party given at the new Gulab Bhawan palace overlooking the Dal lake. (This had recently replaced the Sherghari palace as the Maharaja's residence—it is now the Oberoi Hotel.)

The description makes Hari Singh sound like a movie version of an Indian prince, and in fact a scandalous incident in his life did inspire a film—'Autobiography of a Princess' (1975). In 1924, when he was twenty-nine and about to succeed to the throne of Kashmir, it was revealed that Hari Singh had been the victim of a blackmail plot in Europe. The blackmail had taken place some years before, and it would probably have remained a secret if the blackmailers had not quarrelled among themselves and become involved in a court case during which the story of Hari Singh's part in the drama came out. It had been a classic set-up—a group of conspirators, which included his English ADC, had introduced the prince to a pretty young woman called Mrs Robinson. The prince was very taken with her, and invited her to go to Paris with him. There, in the St James and Albany Hotel, Mrs Robinson had deliberately left the door of their bedroom unlocked and the conspirators had burst in upon the couple demanding thousands of pounds as the price of their silence. The court case in which this story was told took place five years afterwards. Throughout the case the prince was discreetly referred to as 'Mr A', but it was obvious from the evidence that the mysterious Mr A was an Eastern nobleman of power and wealth. Gossip and rumour flew around and finally the prince's name was revealed in the American press. But Hari Singh had been back in Kashmir for five years by then, and the scandal hardly touched him. When his uncle died a few months later he succeeded to the throne without any fuss, and when he paid a state visit to England not long afterwards he was received most politely by King George and Queen Mary.

In Kashmir the Muslim population were far more concerned about the continuing discrimination against them by the maharaja and his government than they were about his private life. Rumblings of discontent grew, and in 1946 a Quit Kashmir movement was launched to pressure the maharaja to abdicate. It was led by Sheikh Abdullah, the son of a shawl-dealer, who pledged himself to liberate his native land from 'Dogra slavery', and it gained such solid support that Sheikh Abdullah was sent to prison.

A year later, at the time of Indian Independence, many Kashmiris had hopes that their region too might become independent, but this was not to be. Instead, there was the sudden raid by tribesmen (described earlier in this chapter) which provoked the maharaja into promising Kashmir to India in exchange for military assistance.

No sooner had that happened than the maharaja, like all the other Indian princes, was obliged to hand over his power to representatives of the people—for independent India was a democracy, and this was no time for princes and feudal fiefdoms. In the case of Kashmir this meant an emergency administration headed by the popular Sheikh Abdullah, fresh out of jail. The maharaja was allowed to remain as token Head of State, or Regent, but Hari Singh declined, and went into exile leaving his son, Karan Singh, in that post. Dr Karan Singh, a distinguished man of letters, went on to a successful career in politics, becoming both Minister for Tourism and Civil Aviation, and then Minister for Health and Family Planning in Indira Gandhi's government.

In 1951 elections were held for the first time in Kashmir's history, and Sheikh Abdullah was confirmed by the people as their choice for prime minister. One of the very first acts of his new government was to abolish the hereditary office of Maharaja, and the Sheikh must have felt that he had at long last achieved his dream of freeing his region. However, events soon showed that what had really happened was that the Dogra dynasty had been replaced by an even greater power, the Indian government itself. It was not long before Sheikh Abdullah was removed from office, and in 1956 the constitution of Kashmir was altered to make the valley an integral part of the Indian Union. There was to be no autonomy, certainly no independence, for Kashmir.

For the next twenty-one years Sheikh Abdullah remained in jail or in exile, but in 1975 he was permitted to become, once again, Kashmir's chief minister—as the region was now part of India it could no longer have a prime minister. The old Lion of Kashmir, as he was known, died in 1983, and his son Farooq Abdullah, a doctor, stepped into his shoes, much as Nehru's daughter Indira Gandhi succeeded him, or as Indira Gandhi's son Rajiv succeeded his mother. Farooq Abdullah, pragmatic and prepared to work with the Indian government, was confirmed as the people's choice in elections held later that year. Since then, however, his popular support has declined, and the Islamic parties that favour independence, annexation with Pakistan, or at least the holding of the referendum, have become ascendant—to the alarm of the Indian government who have now twice suspended the Kashmiri administration and imposed direct rule from Delhi upon the state.

Curfews, rioting, and killing have become commonplace in Kashmir now and it is impossible to predict what might happen in the near future—only one thing seems certain, a happy ending to the troubled story of Kashmir is not yet in sight.

CHAPTER TWO

Travellers' Tales: Descriptions of Kashmir from Earliest Times to the Nineteenth Century

I'll sing thee songs of Araby,
And tales of wild Cashmere,
Wild tales to cheat thee of a sigh,
Or charm thee to a tear.

—W. G. Wills

CHAPTER TWO

Travellers' Tales: Descriptions of Kashmir from Earliest Times to the Nineteenth Century

According to its folklore, the Kashmir valley has welcomed all sorts of famous visitors since, literally, the dawn of creation, when Adam and Eve were there. For, naturally, the valley is said to have been the Garden of Eden, and no one can deny that apples thrive there. In one story King Solomon is supposed to have drained the water out of the valley, in another Moses is reputed to have led his people there from Egypt, and Jesus himself, they say, is buried in Srinagar. His tomb, and his 'footprint' in a rock, are to be found in a back street in the Rozabal area. The legend is that he was taken down from the cross by friends while he was still alive and smuggled out of Jerusalem. From there he made his way to Kashmir where he lived among devoted followers until over the age of a hundred. The most likely explanation for the story is that the tomb at Rozabal was the burial place of a foreigner—perhaps one of the middle-eastern ambassadors to Zain-ul-Abidin's court. His name might have sounded something like 'Jesus', so that over the centuries confusion—and wishful thinking—turned the site into 'Jesus's tomb'.

Stories like these, added to the valley's legendary beauty, have given Kashmir a sort of aura of sanctity, and it has long been a place of pilgrimage where a whole variety of believers have sought their different holy grails. Indeed, it still is. Thousands of Hindus come each year from all over India and beyond to venerate a pillar of ice in the remote Amarnath cave in the north-east of Kashmir, which represents Shiva's lingam, and Muslims flock to the mosque at Hazratbal which houses a hair of the

Prophet brought to the valley from Medina in the seventeenth century.

One of the very earliest descriptions of the country was given by a Buddhist pilgrim from China, Hiuen Tsiang, who travelled there in AD 631. The Kashmiri king of the time was most hospitable to the pilgrim, inviting him to stay in the royal palace, and giving him five attendants and twenty clerks to copy out manuscripts, so it is not surprising that Hiuen Tsiang stayed on in Kashmir for two years. He found the country agriculturally rich, producing 'abundant fruits and flowers' as well as medicinal plants and saffron (for which it is still famed) but, in spite of the king's generosity, the people do not seem to have impressed him. Hiuen Tsiang admired their good looks and their 'love of learning', but for the rest he thought them sadly 'light and frivolous, and of a weak, pusillanimous disposition . . . given to cunning'.

In spite of all these comings and goings, no one in the Western world knew of the existence of Kashmir until comparatively late on in history—more than 100 years after Columbus discovered the New World. It was a Portuguese Jesuit missionary who broke the news, as it were, to Europe.

In 1579 Akbar, curious to learn something about Christianity, had sent an ambassador to the Portuguese mission at Goa with the request that they should 'send me two Fathers, learned in the scriptures, who shall bring with them the principal books of the law, and of the Gospels; for I have a great desire to become acquainted with this law and its perfection.' The Jesuits, naturally enough, had been greatly excited at the idea of a possible convert in the emperor himself, and three priests were dispatched at once to Akbar's court. They were the first of several batches of missionaries sent to the Mughal emperor, for Akbar's enthusiasm for Christianity fluctuated and there were moments when he showed no interest at all and seemed merely to be wasting the priests' time. Then the Father Provincial in Goa would become irritated and recall his men, but each time they left Akbar missed their controversial arguments and discussions and would ask to have them back again; and each time the thought of winning the soul of the most powerful man in Asia would prompt the Father Provincial to relent and try again.

The third mission to Akbar included a Father Jerome Xavier and a Brother Benoist de Goes who arrived at Lahore in 1595. For two years they lived at court fairly uneventfully, until a dreadful accident occurred. In the middle of a magnificent feast being given by the emperor on the terrace of his palace at Lahore, 'fire fell from heaven' and set light to the festive tents and pavilions and burned everything inside them, including a golden throne that the priests estimated was worth a hundred thousand crowns. Worse was to come—the fire spread to the king's palace itself and destroyed not only most of the building, but all Akbar's treasures. It was said that the gold and silver that melted in the heat of the flames ran down the streets of Lahore like streams of water. Akbar decided to leave for Kashmir immediately. Father Xavier and Brother Goes were invited to accompany the emperor, and it is from Xavier's account of that visit, in letters published in Antwerp in 1605, that the western world first learnt that 'the Kingdom of Caximir is one of the pleasantest and most beautiful countries to be found in the whole of India, we may even say in the East.'

Like all its visitors, Father Xavier was struck by the rich greenness of the valley, its pastures, orchards and gardens watered by countless springs and rivers and lakes. But with cruel irony, in the midst of all this seeming fertility, the priest found that the Kashmiri people were starving—the country was in the grip of famine. The shortage of food does not seem to have touched the court at all, but in the city of Srinagar children were being sold by their mothers who, no longer able to provide for them themselves, hoped that someone else might. The priests saw a chance of saving souls and they bought many of the sickest children and baptised them before they died. The emperor's soul, however, remained frustratingly out of their reach. The missionaries had looked forward to the months in Kashmir, believing that in the relaxed and leisurely atmosphere they would be able to spend more time with Akbar and persuade him on to their side, but no sooner had they arrived in the valley than Father Xavier became ill and was out of action for two months, and just as he recovered Akbar himself became sick and, though he received them in his chamber several times, these were not tactful moments in which to discuss conversion.

The disappointed missionaries arrived back in Lahore on 13 November after a tiresome journey during which the elephant carrying their baggage found the mountain paths as difficult as they did, and their only light relief, apparently, was watching the way it used its trunk as a walking stick to lean on.

Akbar died in 1605, his soul still his own, but Father Xavier stayed on at the Mughal court, partly because there was now, of course, the emperor Jahangir to work on, and partly because it was politically useful for the Portuguese to have a foothold at court. He did not return to Goa until 1617, and died there the same year. In the meantime Benoist de Goes had left India in 1603 to make an heroic journey through the heart of Asia to China. After travelling for four years and suffering terrible hardships, de Goes succeeded in reaching the Chinese frontier, but there all his trials and tribulations came to nothing, for he was held prisoner for seventeen months and then he died.

More than half a century was to pass before the glimpse of the valley that Father Xavier had given Europe was expanded into a proper view. But Kashmir's next visitor was one worth waiting for. He was a Frenchman, François Bernier, and though by profession a doctor of medicine, he was also a most entertaining writer—a wit, a gossip, a philosopher, a scientist, an observer—so that one can only echo a letter written by one of his contemporaries who had read his *Travels*: 'Monsieur Bernier is a very gallant man and of a mould I wish all travellers were made of.' It was Bernier's description of Kashmir—'the paradise of the Indies'—that really fired European imaginations and gave Kashmir the aura of glamour that it has retained to this day. Among other things, his descriptions of India and Kashmir and life at the Mughal court inspired Dryden's play *Aurung-Zebe*, a tragedy involving a beautiful captive 'Queen of Cassimere' called

Indamora with whom both Aurangzeb and his father are in love, and though this of course was fiction, it all added a little more mystery and excitement to Kashmir's growing reputation.

François Bernier was born into a farming family in Anjou in 1620. He qualified as a doctor of medicine at Montpelier University, and went to live in Paris. Then he paid a visit to the Middle East which seems to have well and truly whetted his appetite for adventure for, not long afterwards, at the age of thirty-six, he left France and set off on an incredible journey that was to last thirteen years. His objective was to travel down the Red Sea to Abyssinia but on the way he learnt from travellers' talk that Abyssinia had recently become extremely unsafe for foreigners, so at Moka in Yemen Bernier changed direction and took a boat to Surat in India instead, where he arrived in early 1659.

It was a turbulent time in India. Emperor Shah Jahan had been deposed, and his four sons were in the grip of a desperate power struggle for his throne. At the time of Bernier's arrival Dara Shikoh, the eldest son, had just been defeated by his brother Aurangzeb, and now, quite unwittingly—by what he calls 'the strangest chance imaginable'—Bernier himself became involved in the drama.

He was making his way from Surat towards Delhi when he literally ran into Dara Shikoh and the tattered remnants of his army fleeing south to Ahmedabad where they hoped to find refuge and support. The group were in poor shape—in the days of flight since their defeat they had been continually harassed and robbed by the local people. One of Dara Shikoh's wives had been badly wounded in the leg, and when the wretched prince learned that Bernier was a doctor he rather firmly persuaded him to travel with them. Bernier remained with the prince for three gruelling days during which the desperate Dara Shikoh forced them to keep on the move most of the time. 'So insupportable was the heat', writes Bernier, 'and so suffocating the dust, that of the three large oxen of Guzarate which drew my carriage one had died, another was in a dying state, and the third was unable to proceed from fatigue.' But the sorry plight of Bernier's oxen saved him from having to travel further with the doomed Dara Shikoh. For there turned out to be no refuge in Ahmedabad, the city refused to open its gates, and the prince, who by now Bernier tells us was 'more dead than alive', marched his luckless men off again, this time towards Sind in search of an eventual escape to Persia—leaving Bernier marooned with his immobile carriage outside Ahmedabad. Dara Shikoh's followers nearly all died on the terrible march to Sind, and the prince himself was taken prisoner by a local chieftan and callously handed over to Aurangzeb who had him beheaded.

Later, when he came to know Aurangzeb, Bernier judged him to be shrewd but treacherous and 'devoid of that urbanity and engaging presence so much admired in Dara'. For in spite of the chaotic circumstances in which they met, Bernier had warmed to Dara: 'he was courteous in conversation, quick at repartee, polite, and extremely liberal'. Six years afterwards, when Bernier visited Dara Shikoh's beloved Kashmir and saw his lovely garden there, and the fine grey stone mosque he had built, and his school at Pari Mahal with its breathtaking bird's-eye view of the valley, he must have remembered those frantic days of flight and felt a pang of regret for the charming and cultured prince who had crossed his path so strangely.

After more adventures, Bernier finally reached Delhi, but the journey had left him more or less penniless, so that when he was offered a post at the Mughal court he accepted it gratefully. He found himself on the personal staff of Danechmand Khan, one of the emperor's most important nobles, and by reputation the most learned man in Asia. Bernier could not have been luckier with his employer, and he spent the next six years happily at court, not merely acting as physician to Danechmand Khan (and occasionally to the emperor and his ladies as well) but discussing philosophy and religion with his patron, whom he obviously liked and respected, describing to him the latest Western medical discoveries, and translating the works of European thinkers. All the while he was able to indulge his own passionate curiosity. He listened to court gossip, sat in on the emperor's audiences, made an effort to meet foreign ambassadors so that he could learn something of their countries, and explored, investigated, and looked at the sights. Of the 'Tage Mehale' he says: 'this monument deserves much more to be numbered among the wonders of the world than the pyramids of Egypt, those unshapen masses . . .'. As a doctor, Bernier even gained access to that holy of holies, the harem, but these visits were extremely frustrating. An enormous Kashmir shawl would be draped over Bernier's head and an eunuch would lead him by the hand like a blind man—so though Bernier would sense that he was deep inside the harem, he never saw anything and had to content himself with wheedling information out of the eunuchs.

During Bernier's time at court, the most important woman in the harem was Aurangzeb's favourite sister, Roshanara Begum, and because of her pleading Aurangzeb was persuaded to take the court for a holiday to Kashmir in 1664. Unlike his father and grandfather this emperor had no particular love for the place. Indeed in all the forty-nine years of his reign he only went there on this one occasion—but Roshanara Begum claimed to be pining for a change of air, and for her sake he reluctantly agreed to the visit. Bernier suggests that it was not fresh air that Roshanara wanted so much as to come out of the harem and appear in glory 'amid a pompous and magnificent army'—as her sister had done when the court had last travelled to Kashmir with Shah Jahan.

Moving the court to Kashmir was a vastly complicated business, and it is a measure of the love that Jahangir had for the valley that he tolerated the upheaval so many times. For it was not simply a matter of the emperor setting off with his noblemen and a few favourite ladies and their attendants from the harem—though that would have been difficult enough—but the army had to go too, and the bazaars of Delhi followed as well, for, the tradesmen reasoned, what was the point of remaining in a city when all the customers had left. Since each of these groups had families, hangers on, and provisions with them, it was an enormous assembly of people that finally left Delhi with the emperor in December 1664, stirring up the red dust for miles around.

There were at least 100,000 horsemen in the party, as well as thousands of other horses, and camels and elephants and oxen carrying women and children and possessions, and then there were the herds of sheep and goats and cattle that made up a giant walking larder. 'The multitude is prodigious and almost incredible', wrote an astonished Bernier. Indeed, so massive was Aurangzeb's entourage that many of them suspected that they were not going to Kashmir at all, but on their way to besiege the city of Kandahar. And if Roshanara Begum's real ambition was to appear in glory and be the centre of attention, she certainly succeeded—even Bernier was awed at the sight of the princess's procession. She rode out at the head of a line of sixty or more elephants carrying her personal attendants and the important ladies of the harem. Each huge beast was elaborately painted and decorated but Roshanara Begum's was the biggest and by far the most splendid of all, and the gently swaying howdah on its back, in which she sat hidden by silken curtains, was dazzlingly decorated in blue and gold. Close to the princess rode the chief eunuchs, richly dressed and superbly mounted, each one carrying a cane in his hand, and surrounding her elephant came a special troop of female riders from Tartary and Kashmir wearing fantastic costumes. 'Stretch imagination to the limits', writes Bernier, 'and you can conceive no exhibition more grand and imposing . . .'.

But the novelty of even the most spectacular sight wears off, and travelling 'a la Mongole', as Bernier calls it, turned out to be a very slow and tedious business. The emperor would call a halt for days at a time in places where he could indulge his fondness for hunting, so it took the huge travelling city nearly two months to reach Lahore—a journey that could normally be completed in fifteen days.

Bernier had tried to ensure that he at least would be comfortable. He equipped himself with two good horses and a strong camel to carry his baggage, which consisted of a tent, a carpet, a portable bed, a pillow and coverlets, culinary utensils, clothes, crockery and of course food stores—enough for both master and servants, for he had with him his groom, cook, camel-driver and personal attendants. His employer, Danechmand Khan, kindly promised to supply him with a new loaf of bread and a flask of pure Ganges water every day of the journey—each court official travelled with several camels loaded with water—and for these Bernier was especially grateful: 'I am happy at the idea of not being any longer exposed to the danger of eating the bazaar bread of Delhi which is often badly baked and full of sand and dust. I may hope too, for better water than that of the capital, the impurities of which exceed my powers of description . . .'.

The organization that lay behind their vast city-on-the-move greatly impressed Bernier. This was in the hands of the Grand Quarter-Master, who must have been something of a logistical genius. He travelled ahead of the royal party with a complete set of duplicate tents and furnishings so that each time the emperor arrived at a new camping place he found everything in perfect order. It took 60 elephants, 200 camels, 100 mules and 100 men to carry just the royal camping equipment—a far cry from the days of Aurangzeb's nomad ancestors who could take all they needed on the backs of a few horses. The Grand Quarter-Master would select each new site for the camp, which would promptly be cleared and levelled and carefully prepared by Mughal 'engineers', and then he would mark it out. There was a routine layout for the camp, a set place for every class of person in it. The emperor's numerous tents went in the middle, with those of the princesses and the principal ladies around; next came the army, then the royal bazaars that supplied the army, then the noblemen and their attendants. Flagpoles marked each area, and, to show respect, all the tents had to face the emperor's. The noblemen vied with each other to put up the most magnificent canvas palaces, but they had to be tactful about this, for if the emperor thought that anyone's quarters came near to rivalling his own in splendour, he would order theirs to be pulled down.

The royal pavilions were sumptuous inside—lined with beautiful chintzes and satins and exquisite embroideries of silk, silver and gold. The floors were covered with cotton mats three or four inches thick, and on top of these went splendid carpets and rich brocade cushions to lean on, and even the most insignificant of tent poles was intricately painted and gilded. On the outside however they were all identical and rather austere, each one covered in coarse red cloth with wide stripes, but Bernier says that from a distance 'this vast assemblage of red tents, placed in the centre of a numerous army, produces a brilliant effect'.

The camp site was enormous—six or seven miles in circumference—and, in spite of its ordered layout, it was very easy to get lost in it, especially at night when the dust and the smoke from thousands of little cooking fires, burning camel and cow dung, stung the eyes and made it difficult to see anything. And then there was the added hazard of the strung-out guy ropes of all the hundreds of tents which lay in wait to trip up the unwary traveller as he blundered about in the dark. Thieves seem to have flourished in the camp too, and people who did get lost would make their way as hastily as possible towards the camp landmark called the *aquacy-die*, a tall mast-like pole near the royal tents which had a lantern burning at the top. There they could sleep the night protected from robbers, or get their bearings and try to find their own tents again.

But on the whole, life in camp was tolerable and Bernier had nothing worse than boredom to complain about—at least until the march from Lahore to the town of Bimber in the foothills of the mountains separating India from Kashmir. The emperor and his enormous entourage had had to wait in Lahore for two months for the snow on the high passes to melt, and by now it was April and the burning heat of the Indian summer was relentless. Bernier's letters become increasingly desperate with each step of that journey. Four days from Lahore he writes: 'I declare, without the least exaggeration, that I have been reduced by the intenseness of the heat to the last extremity; scarcely believing, when I rose in the morning, that I should outlive the day'. Six days from Lahore: 'Every day is more insupportable than the preceding...'. At this point the huge party had to cross the Chenab river on a bridge of boats and things went badly wrong. In the confusion, one of Danechmand Khan's camels was swept away and, as luck would have it, it was the camel carrying the precious bread oven—'I fear I shall be reduced to the necessity of eating the bazaar bread', Bernier observes gloomily. Eight days from Lahore: 'what can induce a European to expose himself to such terrible heat, and to these harassing and perilous marches...my life is placed in continual jeopardy.' Ten days from Lahore:

The sun is but just rising, yet the heat is insupportable. There is not a cloud to be seen nor a breath of air to be felt.... The whole of my face, my feet and my hands are flayed. My body too is entirely covered with small red blisters which prick like needles.... Heaven bless you: the ink dries at the end of my pen, the pen itself drops from my hand....

Happily, just at the point when Bernier—and presumably his thousands of travelling companions—felt they could endure no more, they reached Bimber and the foothills: 'God be praised', writes a new Bernier, 'the atmosphere is evidently cooler, my appetite is restored, my strength improved...'. This relief was short-lived for all but a privileged few, for Kashmir was too small a country to support many thousands of visitors and at Bimber the emperor's entourage was divided up into two groups: a comparatively small party was selected to accompany him into the valley, while the vast majority of the travellers was left behind to languish at Bimber or Lahore and wait for the emperor to return from his holiday in the mountains. Guards were set on the mountain passes to ensure that no unauthorized person slipped through to the valley of Kashmir, and the groups who were to go with Aurangzeb were ordered to stagger their departure so that the difficult path through the mountains would not become overcrowded—a sensible move in view of the fact that the chosen few had 30,000 porters to carry their luggage across the mountains. The king alone employed 6000; Bernier hired three.

The march from Bimber to Srinagar took five days and it should have been a pleasant one, with everyone delighting in the cool air and the novelty of the scenery and the beauty of the brilliant spring flowers all around, but an accident occurred which, Bernier writes, 'dampened all our pleasure'. It happened when the emperor was crossing the highest of the mountain passes, the Pir Panjal. Along the mountain track behind him stretched the long line of elephants in single file carrying Roshanara Begum and the ladies of the harem. Quite suddenly, one of these elephants stopped and stepped backwards and collided with the elephant behind, this one tried to back away, only to crash into the elephant behind *it*, and so on, all down the line. The track was too narrow for the frightened elephants to turn or pass each other, and in the ensuing panic fifteen of them fell over the precipice at the edge of the path. Only three or four of the women were killed, but there was no way of lifting or saving any of the elephants and the fifteen poor beasts were left to die. Bernier came along the track two days after the accident and reported that some of the elephants were still alive and waving their trunks pitifully. In an early edition of Bernier's *Travels*, published in Amsterdam in 1672, there is a map of Kashmir on which the Pir Panjal pass is marked by a touching little pile of dead elephants.

After all the months it had taken to reach Kashmir, one might have expected the place to be something of an anti-climax, but far from it. Bernier was captivated by the country and the tedium of the journey was quite forgotten in his delight at the magnificent mountain scenery all around, and the prettiness of the valley: 'the whole kingdom wears the appearance of a fertile and highly cultivated garden', he says. He marvels at the European flowers and fruits that grow: apples, pears, plums, apricots, walnuts, melons, watermelons and most herbs, and observes wryly that the lower slopes of the hills are so crowded with cows and goats and bees that Kashmir 'could literally be said to be flowing with milk and honey'.

It was not only the natural beauty of the country that attracted Bernier. He was drawn to the people too: 'The Kachemirys are celebrated for wit, and considered much more intelligent and ingenious than the Indians. In powetry and the sciences they are not inferior to the Persians . . .'. On Aurangzeb's arrival

1. The ladies terrace at the back of Nishat garden photographed in 1900 by Geoffroy Millais. This pavilion no longer exists, and neither do the 'pillared porticos' which once shaded the stone-benches across the central canal. But the Mughal retaining wall with its arched niches and octagonal gazebos at either end (not seen in this photograph) remains.

2. The black marble pavilion in Shalimar Bagh, *c.*1900.

3. A houseboat called 'Kismet', rented by an English family, *c.*1900.

4. The Nishat garden seen from across the water in 1900. Today a road separates the garden from the lake, and the waterside pavilion has been demolished.

5. Char Chenar island from a drawing of 1835 by Godfrey Vigne. The pretty Mughal pavilion is clearly visible.

6. The same view of Char Chenar island today, with no trace of the pavilion.

7. The simple but elegantly proportioned Pather Masjid mosque in Srinagar, built by Nur Jahan.

8. A view of the Akhund Mullah mosque in winter. This exquisite little mosque built by Dara Shikoh in the mid-17th century is in desperate need of restoration and care.

9. One of the two gateways to Akbar's city of Nagar-Nagar which can still be found. Painted by Captain Molyneux, 1917.

10. Partially submerged Hindu temple in Manasbal lake. Watercolour, 1886, by Charles Cramer-Roberts.

11. East view of the sun temple, Martand, painting by Charles Cramer-Roberts, 1886.

12. A Kashmiri papier mâché merchant at the turn of the century, surrounded by his wares.

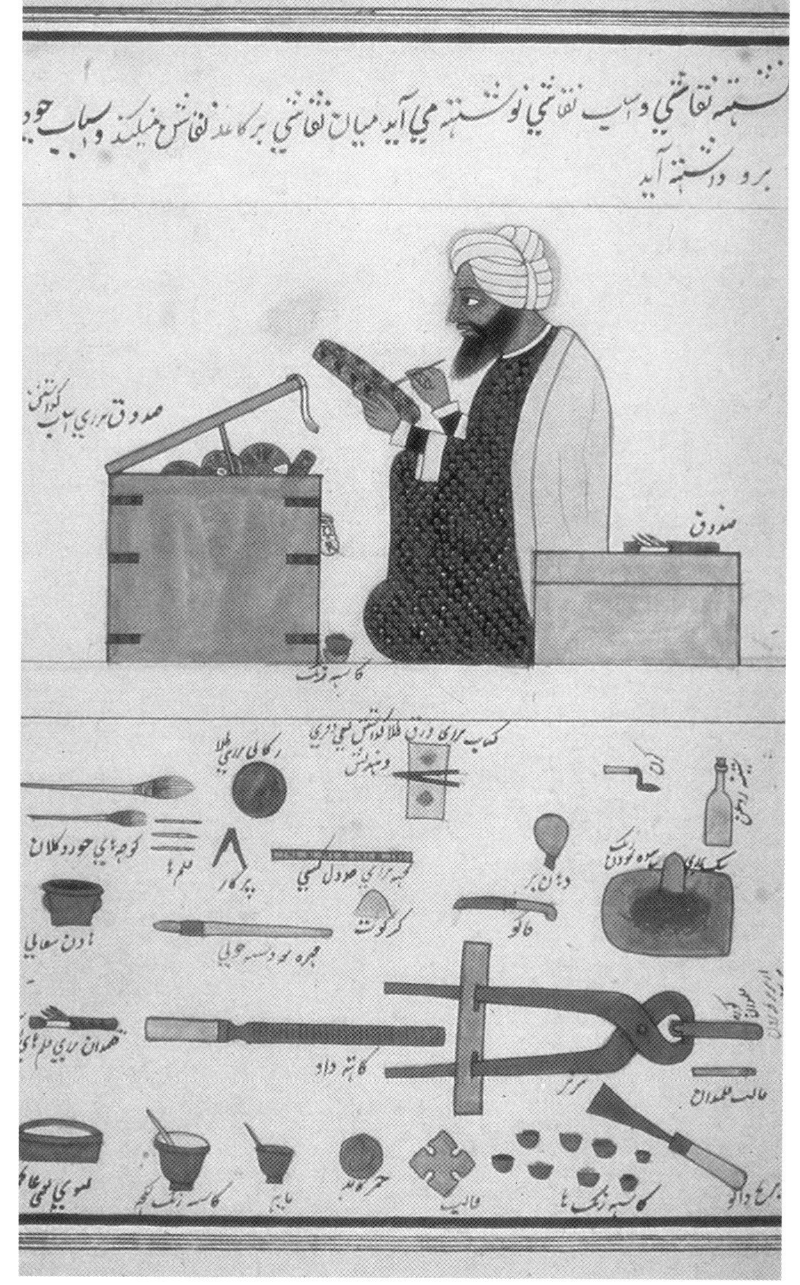

13. A Kashmiri papier mâché painter at work, his tools displayed below. Nothing has changed since this drawing was done in about 1860.

14. Kashmiri shawls being washed on a river bank, a painting commissioned for the Paris Universal Exhibition of 1867.

local Kashmiri poets had competed with the Mughal court poets to produce the most magnificent welcoming verses for the emperor. The results were such exaggerations and extremes of 'extravagant hyperbole' that Danechmand Khan and Bernier were reduced to irreverent and un-courtier-like laughter.

Bernier admired the local architecture as well, for though the houses were mostly made of wood they were two or three storeys high and 'well built', and even in the city they had flower gardens: 'and many have a canal, on which the owner keeps a pleasure-boat, thus communicating with the lake'. The suburbs of Srinagar were even more attractive: 'Most of the houses along the banks of the river have little gardens, which produce a very pretty effect, especially in the spring and summer, when many parties of pleasure take place on the water'. At the foot of Hari Parbat hill stood the 'handsome houses' of Akbar's city where the Mughal court stayed when they were not camping or picnicking in the royal gardens; the foothills around the Dal lake were 'crowded with houses and flower gardens'; the islands in the lake were 'so many pleasure grounds ... beautiful and green in the midst of the water'; and finally, there were the royal gardens themselves, 'laid out with regular trellised walks', their open pavilions surrounded by shady trees and canals and pools and fountains. No wonder Bernier writes: 'I am charmed with Kachemire. In truth, the kingdom surpasses in beauty all that my warm imagination had anticipated. It is not indeed without reason that the Mogols call Kachemire the terrestrial paradies of the Indies ...'.

These glimpses of the valley are valuable to us, for Bernier was one of the very few Westerners to see and write about Kashmir under the Mughals—a Kashmir graced and protected by the love of successive great emperors in Agra and Delhi. Of course it was not Utopia. There was poverty, even famine (as Father Xavier had witnessed), but there was a degree of order, freedom from oppression and religious tolerance under the Mughals that Kashmir would not know again for more than 200 years.

It is tragic to compare Bernier's picture of the lovely valley and its witty ingenious people with those of later travellers, such as Forster and Moorcroft, who visited Kashmir under Afghan and Sikh rule. Where Bernier saw delightful flower gardens, free-flowing waterways and handsome houses, they saw only decay: ruined pavilions, tumbledown houses, choked canals, overgrown gardens and filthy, unattractive people whose natural shrewdness had been distorted into low cunning in order to survive.

Danechmand Khan, being of a scientific mind, dispatched Bernier all around the valley to investigate various phenomena. In particular, Bernier was told to check up on the miraculous cures being reported by the eleven mullahs of a certain mosque, who also claimed to be able to lift a heavy stone by merely resting their fingertips on it. The cures, Bernier discovered, were easily explained by the fact that mullahs were giving free meals to all who 'miraculously' became well; consequently the place was crammed with dozens of people feigning sickness. Then Bernier asked for a demonstration of the mysterious stone-lifting and noticed that though the mullahs all agreed that the rock was as light as a feather, they were actually straining every muscle to lift it, and were using their *thumbs* as well as their forefingers. But by now the mullahs were looking at Bernier somewhat threateningly, so he decided it would be expedient to leave.

Bernier was sent north to inspect the Wular lake, and he travelled south on a route that was to become a favourite with travellers right up to the present time, for it takes in what Bernier called the 'ancient idol temples in ruins' at Avantipur and Martand, as well as the Mughal gardens at Bijbihar, Achabal and Verinag—nowadays only pale shadows of what they were like in Bernier's day, when they were still enjoyed and cared for by the Mughal court. Bernier liked Achabal best, with its ice-cold gushing spring and waterfall, but at Verinag he was amused to see that some of the larger fish still had the gold rings through their gills with which Nur Jahan had had them decorated two generations before. Bernier had intended to travel on and visit the famous holy cave at Amarnath next, but the idea had to be abandoned. Danechmand Khan had become restless for his company, and sent word that he should return at once to Srinagar. Back in Srinagar Bernier amused himself trying to catch glimpses of upper-class Kashmiri women without their veils. The good looks of the ordinary women of the town made him long to see what their more sheltered, shut-away sisters looked like. First he tried a little trick that he learnt in India—whenever a richly harnessed elephant passed through the streets he followed it, knowing that when the ladies indoors heard the tinkling of the elephant's bells they would not be able to resist rushing to the window and peeking out. But then an old Kashmiri offered to make it easier for him. Pretending that Bernier was a relative, 'rich and eager to marry', he took him on family visits and, naturally, all the women and girls of the various households came to look him over—and share the sweets he brought. Bernier was thus able to conclude 'that there are as handsome faces in Kachmeire as in any part of Europe.'

The Mughal court returned to Delhi in the autumn of 1665, and soon afterwards Bernier left Danechmand Khan's service and set off to explore Bengal with the famous French traveller Tavernier—who was on his sixth visit to India at the time. Then, in 1667, eight years since he had first disembarked in India, Bernier was back in the port of Surat, waiting for a boat to leave. There, in rather grisly circumstances, he came across another well-known traveller, John Chardin, who was in India buying diamonds (he later emigrated to England and became court jeweller to Charles II). They met whilst watching a young Hindu widow being burned alive on her husband's funeral pyre. Unlike most of the *satis* that Bernier seems to have been irresistibly drawn to watch, this girl went to her death willingly, with a look of 'brutish boldness or ferocious gaiety' on her face that horrified the two Europeans.

Two years later—for he lingered some time in Persia on the way home—Bernier arrived back in Paris, and the following year, in 1670, he published his *Travels*, which was greeted with great enthusiasm. He died in 1688 and was buried, appropriately enough, by his friend, D'Herbelot, a distinguished Orientalist whose dictionary of eastern knowledge was the source for much of the epic poem about Kashmir, 'Lalla Rookh'.

After Bernier's delightful descriptions of the country a silence descended. Nothing more was heard in Europe about Kashmir for nearly fifty years. But then another of those intrepid Portuguese missionaries arrived in the valley. Father Desideri only intended to pass *through* Kashmir, for he was on his way from Goa to Lhasa, but no sooner had he arrived, in November 1714, than it began to snow so thickly that his route to Tibet was blocked and he was obliged to remain in the valley for the winter. The poor priest was plagued with dysentery, so he was not altogether unhappy at the idea of staying in one place for a while. He took a house in the city for six months, presented letters of introduction from the court in Delhi to the governor, who treated him 'with honour', and he continued his Persian studies. Father Desideri clearly found Srinagar and its surroundings as peaceful and pretty as Bernier had done:

> A big river flows through the middle of the city, and near by are large and beautiful lakes, whereon with much pleasure and amusement one can sail in small boats or in well-found larger vessels. A great many delightful gardens near or on the borders of these lakes form, as it were, a most ornamental garland round the city, which contains splendid buildings and well-laid streets, squares and bridges. The houses of merchants and of common people, and also some of the noble, are built of stone and brick, but outside they are of well and diligently carved timber.... The whole district round the city is not only beautiful, but extremely fertile, producing abundant crops, and a great variety of fruit.... In spring, the many European flowers, which are not found in quantities in other parts of Mogol, such as roses, tulips, anemones, narcissi, hyacinths and the like are a great delight to the people of the country.... For these reasons Kascimir is called by everyone in Mogol *Beheset*, which means 'Terrestrial Paradise'.

In May when the snow melted and the road to Tibet was once more passable, Father Desideri resumed his journey to Lhasa. Leaving the earthly paradise of Kashmir, he toiled on into neighbouring Ladakh, and the contrast between the two must have been a shock, for he describes that country in no uncertain words as 'mountainous, sterile and altogether horrible'.

Little would the good priest have believed that Kashmir's days as paradise were running out, that he was the last European to describe the valley in its prime, and that future travellers were more likely to talk about Kashmir in the sort of words he had used for Ladakh—'altogether horrible'.

After Aurangzeb's death and the break-up of the Mughal empire, it became increasingly difficult to get to Kashmir from India. The Royal Route from Delhi—the one that Bernier had travelled on with the Mughal court—lay across the Punjab, which was now a battleground between the Sikhs and the Afghans, and far too dangerous to be risked by merchants and travellers. But traders are ever-resourceful, and the merchants took their caravans to and from Kashmir by a route that began further east and led through the foothills of the Himalayas and up into Kashmir through the state of Jammu. It was a longer and far more wearying way to go, but a comforting, protective barrier of mountains lay between it and the warring armies and marauding deserters of the Punjab. However, even the new route was not safe for long, for the tiny mountain kingdoms through which it passed were affected by the general instability and began to quarrel and fight with one another, making this, too, a dangerous part of the world. Even if a man got through to Kashmir physically unscathed, he could not hope to arrive with all his possessions intact, for every little state had set up a customs house and taxed the merchants passing through in an arbitrary way. The better off you looked, the more you paid. There were no less than thirty customs posts between Kashmir and Lucknow, and then of course there was always the danger of highwaymen and brigands who frequently stripped a traveller of all he had, leaving him literally naked on the path.

Not surprisingly, no European attempted this journey for years, until in 1783 George Forster, a civil servant with the East India Company who had lived in Madras for twenty years, decided to make his way home to England via Kashmir, Kabul, the Caspian Sea and St Petersburg. Forster's account of this astonishing and perilous journey is fascinating to read because though he means to be strict with himself and describe the more impersonal things like, say, the geography of a country, somehow he gets sidetracked into how sore his feet are, or whether swallowing an Indian fly makes him feel sick because it has poison in it, or because it is buzzing round his stomach, or how he has caught fleas from an old lady sharing his camel whose hair was 'a complicated maze of filth, which had never, I believe, been explored by comb . . .'.

He travelled alone, sometimes on foot, sometimes on horse-back. For quite long periods he rented a box-like seat slung on a camel (this was the most uncomfortable way, because the box was always too small). Whenever he could, he linked up with merchants' caravans for protection, and now and again he would hire a servant to help with the tiresome business of foraging for food, fuel and water each evening, as well as rubbing down the horse. He carried very little in the way of possessions—a pipe to smoke in the evenings, a thin mattress, a quilt, a canvas bag with utensils, a few stores and some spare clothing, and a large black blanket that could be thrown over a light bamboo frame to serve as a small tent. 'Nor should any person travelling in my manner', warned Forster, 'have more equipage—a larger will raise unfavourable conjectures, and subject him to frequent investigation, delay and taxes.'

In that wild part of the country, and in those unruly times, it was essential for Forster to travel in disguise. Somehow he had to account for his white skin, so the tale he put about was that he had been born in Turkey and raised in India, where he had served in the army until he grew tired of it, and now worked as a travelling merchant. 'The story', he writes—'not very complex, possessed plausibility sufficient to procure common belief, and I myself had entered so warmly into its spirit, that I began to believe it'.

Very little is known about George Forster's life. Even his birth date is not recorded, but he comes through the pages of his book as such an amiable, good-humoured man that one can't help liking him and fearing for him as he travels the risky road to Kashmir, at every moment in danger of being exposed as an infidel impostor and killed. He kept a daily diary and this aroused the curiosity of his travelling companions on the road who demanded to know what it was that he was so busy writing. Forster reassured them that he liked to keep his day-to-day expenses accounted for—but then one of the more sharp-eyed and suspicious of them asked why he wrote from left to right in the European way, and not from right to left. Forster explained coolly that he was writing in *Turkish* and that Turkish went from left to right—quite untrue at that time, but it seemed to satisfy them. More serious was the occasion they caught him urinating *standing up* instead of squatting, as a true Muslim would. Inventing rapidly, he explained that he had picked up this unholy habit in the army—soldiers, he told them apologetically, always preferred to stand for reasons of speed and safety.

Forster's luck held. His disguise was never penetrated, their caravan was not attacked by brigands, and he reached Kashmir unscathed in the spring of 1783. By now the Afghans had governed the valley for thirty odd years, but Forster's is the only first-hand account by a Westerner of what the country was like under their rule. He spent his first night in Kashmir miserably, in a boat on the Jhelum river in the pouring rain, and decided that the only reason he did not catch his death of cold was because smoking his pipe protected him from 'noxious vapours'. It is ironic, in these days of Government Health Warnings, to think that Forster never doubted that the excellent health he enjoyed throughout his journey was entirely due 'to the common habit of smoking tobacco'. In spite of this bad start, the sight of the Kashmir valley completely bewitched him. It was springtime and the red and white roses were blooming, the fruit trees were in flower everywhere, and Forster wrote that he felt as though he were standing in 'a province of fairyland'.

But as their boat moved down the Jhelum river towards the capital, Srinagar, rumours of the tryanny of the Afghan governor began to reach them, and a different picture of the valley emerged. The governor, Azad Khan, was an adolescent of eighteen who had only recently inherited the post from his father, but in the first three months of his reign he had become 'an object of such terror to the Kashmirians, that the casual mention of his name produced an instant horror and an involuntary supplication of the aid of their prophet'. The father had been bad enough—notorious for his 'wanton cruelties and insatiable avarice' he liked to punish even trivial offenders by tying them back to back and throwing them into the river to drown. He plundered the people's property and violated their women, but, Forster wrote, 'they say he was a systematical tyrant', whereas the really frightening thing about Azad Khan was that he was wildly capricious as well as cruel and with a ferocious temper 'rarely seen in the nature of man'. Azad Khan slit his doctor's stomach open when he failed to produce an instant cure for an eye disease, murdered his own wife, and once amused himself taking pot shots with his musket at the crowd which had gathered to watch him pass. All the same, whilst appalled at the way they were treated, Forster could not bring himself to like the Kashmiris very much. He confessed this rather guiltily to a Georgian merchant who had lived for a long time in Kashmir. This man had known the valley under kinder governors, and he assured Forster that it was only since the oppressive rule of Azad Khan's father that the people had become 'dispirited, their way of living mean, their dress slovenly'.

The Kashmiri who irritated Forster most was his own landlord. He had been particularly recommended to find lodgings in Srinagar with this man, a merchant, but unfortunately the merchant had been told that his 'Turkish' guest was rich, and from the moment he arrived he fawned on him in an obsequious way, smothering him with what Forster called disgusting attentions. 'He commenced with embracing my legs, and ended in washing my beard in rose water', says Forster shudderingly. The problem solved itself when another, genuinely rich, merchant arrived from Constantinople—the host switched all his attention to him and from then on, Forster reports wryly, 'the Sheikh had no leisure to say a civil word to anyone'.

By the time of Forster's visit to Kashmir, apparently the only relic of Mughal times left in a reasonable state of repair was the Shalimar garden, 'which is preserved in good order and is often visited by the Governor . . .'. As for the rest, Forster says sadly that Kashmir was testimony to the barbarity of the Afghan nation who 'have suffered its elegant structures to crumble into ruins . . .'. The city of Srinagar, which looked so beautiful from a distance with the earth roofs of its houses a colourful mass of spring flowers (Forster thought the effect was like looking down on the beds of a formal garden), was, on closer inspection, 'choked with the filth of the inhabitants'.

Two or three weeks in Afghan Kashmir were quite long enough for Forster, and—still of course in his guise as a Turkish trader—he applied to the governor for permission to leave, for no one was allowed to travel out of the valley without written authorization. To his horror this was refused. Azad Khan decided that he should be recruited into the Kashmiri army instead, for Turks had a fine fighting reputation. This was the beginning of a nightmare for Forster, but luckily he had not been near the court himself, so he lay low and tried to present another petition to leave through a sympathetic banker friend he had made, this time using another name and declaring himself to be a Muslim trader. As bad luck would have it, the poor banker fell out of favour at court on the very day he was to present the petition (he was later killed by Azad Khan), and Forster became more and more alarmed: 'The obstacles that stood in the way of my departure now became serious, and gave me much anxiety. I was thrown into the power of a capricious tyrant . . .'.

Risking his own favour at court, if not his life itself, the Georgian trader with whom Forster had become friendly agreed to present the petition. It was a cliff-hanging affair. The trader was obliged to appear at court for fifteen successive days before the governor heard the petition and gave permission for the 'Muslim merchant' to leave. Then Forster wasted no time. He hired a horse and, six weeks after his damp arrival by boat, left Srinagar behind as quickly as he could. But the drama was not yet over. Two nights later, as he slept, his coat was stolen, and tucked in the pocket was his precious permit. Forster blamed himself for this—it was 'some spark of vanity' that had made him buy a *red* coat, and had it not been such a 'gaudy garment' it would undoubtedly not have been stolen. Nothing on earth could have persuaded Forster to go back to Srinagar, where no doubt Azad Khan's men were already searching for the Turk he wanted in his army. Instead, with his heart in his mouth, he offered the border guards a large bribe, and to his intense relief they allowed him to pass across. Forster was to suffer many more uncomfortable and dangerous moments on his long journey home, but nothing was ever as tense and nerve-racking as his escape from Kashmir.

He arrived in England in July the following year, but later returned to India by a more conventional route. In 1790 the first lively volume of his travels was published, but two years later he died at Nagpore (Nagpur) in India. Happily, a second volume was put together from his papers by an unknown editor and published posthumously in 1798.

In the years that spanned the end of the eighteenth century and the beginning of the nineteenth, India seemed to be always in the news in Europe. In London the trial of Warren Hastings, on charges of misuse of power and corruption whilst Governor-General in India, had ended in his acquittal in 1795, but the proceedings had lasted for no less than seven years and at times the evidence brought the splendour and squalor and colour and excitement of remote India vividly to life, with its tales of rajahs and princesses and eunuchs and treasure and jewels and intrigue. 'Every step in the proceedings', writes Macaulay, describing the opening days of the trial, 'carried the mind...far away, over boundless seas and deserts, to dusky nations living under strange stars, worshipping strange gods, and writing strange characters from right to left.'

Out in India itself, the East India Company was battling successfully to establish Britain as the supreme power in the land, and returning Company men like George Forster were astounding the English at home with their exotic stories and often fabulous possessions. It was not surprising that India soon became all the rage. The Prince of Wales himself was among the first enthusiasts. In 1815 he commissioned the architect John Nash to transform the Royal Pavilion in Brighton into a fabulous Oriental palace. Other grand houses had been built or redecorated in Indian style, and minarets and Mughal arches sprouted on all sorts of solemn and unlikely buildings. Most extraordinary of all was the sweeping fashion for the Kashmiri shawl: no lady who considered herself *à la mode* could possibly be without at least one of these, and preferably several. The Empress Josephine is supposed to have owned 200. This curious passion for an item of clothing, which in its native land was worn by *men*, was to last for 100 years—probably the most enduring 'craze' there has ever been in the fickle world of fashion.

All this meant that in 1817 the public were in just the right receptive mood for 'Lalla Rookh', an epic poem about India and Kashmir written by the popular poet Thomas Moore. Presumably that is what Longman, his publishers, were banking on, for they paid Moore the huge sum of 3000 guineas before one word of the poem was written. The author himself was so worried about their extravagance that he wrote anxiously to enquire if they were *sure* they knew what they were doing. They did—'Lalla Rookh' sold 83,500 copies and went into fifty-five editions before Longman's copyright expired in 1880.

The poem became a cult; even the Russian royal family loved it—so much so that when the Grand Duke Nicholas (soon to be Tsar) and his wife paid a visit to Berlin in 1822, a splendid entertainment was organized during which the royal party and their hosts acted out 'Lalla Rookh' in a series of lavish *tableaux vivants*. The Grand Duchess, of course, took the part of Lalla Rookh, and the Grand Duke played the King of Bucharia, and all agreed that it was a delightful, fairy-tale evening.

A side effect of the immense popularity of the poem was that it indelibly glamorized Kashmir's reputation in the eyes of the public. The ironic part of the whole story is that Dublin-born Thomas Moore had never been near the East in his life, let alone set foot in the Kashmir valley, though long afterwards there was talk of erecting a statue of him there. No wonder the compliments that pleased him best were the ones from travellers who professed to be amazed that he could so accurately describe countries he had never been to. One well-travelled Colonel remarked wryly, 'Well, that shows me that reading over D'Herbelot is as good as riding on the back of a camel'—D'Herbelot's dictionary of oriental knowledge had been an important source of reference for Moore, who also drew on Bernier's and Forster's descriptions of Kashmir.

'Lalla Rookh' tells the (fictional) story of the beautiful daughter of Emperor Aurangzeb who has to travel from Delhi to Kashmir for her arranged marriage to the king of Bucharia. The description of Lalla Rookh's departure from Delhi on a magnificent elephant is hardly different to Bernier's of Roshanara Begum. The princess is bored on the long journey, and in an effort to amuse her a handsome young Kashmiri poet, who happens to be travelling with them, is asked to come and entertain her with his poems and stories. Naturally, she falls in love with him, and as each day of the journey brings her closer to Kashmir and her marriage, she becomes thinner and paler and sadder, but still determined to do her duty. At long last the party reaches the 'paradise of the Indies' and Lalla Rookh is taken by barge to the Shalimar garden where her wedding will take place. Heavy-hearted, she staggers from the barge to meet the king of Bucharia, her bethrothed, and as soon as she sees him she falls in a faint, for he is of course none other than the Kashmiri poet (the king had disguised himself so as to see what she was *really* like), and they live happily ever after.

The whole of the poem oozes romance, and the verses on Kashmir are a song of praise that could have been specially commissioned by its tourist office:

> Who has not heard of the vale of Cashmere,
> With its roses the brightest that earth ever gave,
> Its temples, and grottos, and fountains as clear
> As the love-lighted eyes that hang over their wave?

or:

> If woman can make the worst wilderness dear
> Think, think what a heaven she must make of Kashmir.

Since Moore's descriptions of real places like the Shalimar garden, the Isle of Chenars and the Dal lake are all mixed up with his exotic flights of fancy, it is not hard to see how, in peoples' minds, Kashmir came to be a sort of eastern Camelot. A great deal has been written about the country since 'Lalla Rookh' was published, much of it critical and unsparing of the place and its people, but Moore's rosy picture has somehow endured all the batterings it has received and, without realizing it, even today's visitors to Kashmir (who are most unlikely to have read his poem) probably have their expectations heightened and coloured by the flowery imagination of this Irishman who never went there at all.

'Lalla Rookh' was published while Kashmir was still cut off from the world by the difficulties of getting to the valley and the dangers of tangling with her despotic Afghan rulers if you did. In 1817 she was just the place—secret, mysterious, isolated—around which a romantic poet like Moore could weave his dreams and let his fantasies run wild. But two years later Kashmir was wrested from the Afghans by the Sikhs, and joined the world again. It became possible for a European to obtain permission to visit the valley from the Sikh leader Ranjit Singh—though if he had known how vehemently his visitors would condemn his rule there he would surely not have been so co-operative.

The first European known to have taken advantage of this new freedom was William Moorcroft, a veterinary surgeon with the East India Company. Moorcroft was a Lancashire man who had begun training in Liverpool as a surgeon, but was persuaded to switch to veterinary work when a virulent disease attacked and decimated the local livestock population. Later he moved to London where he ran a successful veterinary practice, but Moorcroft was a man of impulsive enthusiasms, and one of these—some scheme for manufacturing horse shoes—lost him his savings. At this point he was approached by the East India Company who offered him a job as superintendent of their military stud in Bengal. He signed on and arrived in India in 1808, and seems to have done his job efficiently. For every ten horses that were sickly when he took over, he wrote a few years later, there was now only *one*. But he was convinced that the native cavalry horse of India could only really be improved by cross-breeding it with the sturdy horses of Central Asia, and that a number of these should be collected for this purpose. Moorcroft's enthusiasms went beyond his veterinary work: he also fervently believed that Britain could set up a lucrative trade, particularly in woollen cloth, in the Himalaya and beyond, and that this possibility should be explored; and he felt strongly that if only British manufacturers were supplied with detailed information on how the coveted Kashmiri shawls were made, they could beat the rest of the world at shawl production.

With all these ends in view, Moorcroft set off in 1819 on an immense trek that was to take in Ladakh, Kabul and Bokhara. He took with him George Trebeck, the young son of a solicitor friend who had begged to join the expedition as artist and map-maker, and Mr Guthrie, an Indian doctor. To start with they also had an eminent geologist, but he behaved so badly towards the servants and porters that he was asked to leave.

The East India Company, while granting him permission to make the expedition, did not seem to share Moorcroft's excitement over these projects. He had already made one expedition to Tibet to collect shawl-wool goats for breeding in England, and perhaps they thought that was enough. The general feeling among the British in India seemed to be that Moorcroft's new expedition was 'an undertaking from which little, except the gratification of his own taste for a wandering life, was to be expected'. This was most unfair: Moorcroft was to prove one of the most diligent, most painstaking and stoic travellers of his time—indeed it was his dogged determination to fulfil his mission and bring back those Central Asian horses against all odds that led, in the end, to his death. The only aspect of his journey, apart from the very long time it took, that the Company could legitimately complain about was Moorcroft's dabbling in the politics of Ladakh. Moorcroft and his companions spent two years in that country (from 1820 to 1822) and he formed the view—it was another of those enthusiasms—that the place would be far better off under the British. The Raja of Ladakh agreed, and, using Moorcroft as a willing go-between, he offered his allegiance to the British. The East India Company was extremely annoyed. It stopped Moorcroft's pay and coldly declined the raja's offer to take over his country.

Happily for the expedition, this little excursion into politics does not seem to have worried anyone else, for when Moorcroft's caravan of 70 people and accompanying horses, sheep and goats arrived in Kashmir over the mountains from Ladakh, they were welcomed at Sonamarg by a representative of the Sikh governor of the country, and as their party approached the city of Srinagar, a detachment of soldiers was sent to escort them to their camping place. The houses and streets of the city were crammed with people trying to catch a glimpse of the travellers, but as Moorcroft's party forced a path through them it became obvious that the crowds were there out of more than idle curiosity: 'hands were stretched out, and cries addressed to us, praying for our interference to save the inhabitants from starvation'—an inhuman order had recently come from Ranjit Singh forbidding the people to sell any of their rice crop until a deficit in the preceding year's revenue from Kashmir had been paid. Moorcroft and his companions were allocated a dilapidated house in an overgrown garden that had been created by one of Jahangir's governors of Kashmir, Dilawar Khan. Though it was shabby and neglected, its convenient location in the heart of the city, and its pretty views across the lake-like waterway (known as the Babadamb) towards the Shankaracharya hill, made it an attractive proposition, and Moorcroft took up his headquarters in a summerhouse overlooking the water. (A school called the M. L. Higher Secondary Institute occupies most of the site today, but the views are still pleasant and three old chenar trees prove that this was once a Mughal garden.) The wretched crowd that followed them was eventually persuaded to go away, but Moorcroft and his party seem to have lived in a state of semi-siege, for there were always dozens of petitioners, beggars, patients—as well as the merely curious—squatting around their camp.

It was now November 1822, and they planned to stay in Kashmir until the following spring or summer. Moorcroft put aside one day a week—every Friday—for the treatment of the sick. In Ladakh he recorded in his diary that he had performed fifty cataract operations in two months. At one time in Kashmir he had as many as 6800 patients on his list—a large proportion of whom were suffering from 'the most loathsome diseases, brought on by scant and unwholesome food, dark, damp, and ill-ventilated lodgings, excessive dirtiness, and gross immorality'. The remainder of each week was devoted to research, first and foremost of course to that detailed account of shawl production in Kashmir that Moorcroft believed would be so helpful to British manufacturers, but after that he investigated and wrote about every aspect of life in Kashmir—from its agriculture, including the famous floating vegetable gardens, to its history, economy and the character of its people. Everything, that is, except its sexual habits—though in view of the sarcastic comments Victor Jacquemont made when he arrived in Kashmir a few years later, it seems obvious that Moorcroft thoroughly explored this aspect too, but for his own reasons omitted to write about it. Moorcroft collected shawls, shawl designs, samples of wool, seeds—anything he thought might be useful to the Company. His work in Kashmir was down-to-earth and scientific, and as far removed from the baroque verses of 'Lalla Rookh' as a tractor is from Cinderella's coach and horses. But the sad part is that it was 'Lalla Rookh' that people remembered best, not Moorcroft.

Moorcroft and his companions made only one excursion into

the countryside that winter, but they explored the city itself and its immediate surroundings and found them depressing. Since Forster's visit the Afghans had built themselves a fort right on the top of Hari Parbat hill, but apart from that nothing had changed except to deteriorate. Moorcroft described the city as 'a confused mass of ill-favoured buildings, forming a complicated labyrinth of narrow and dirty lanes, scarcely broad enough for a single cart to pass, badly paved, and having a small gutter in the centre full of filth, banked up on each side by a border of mire.' Those large well-built houses that Bernier had liked so much with their once-picturesque earthen roofs were now in a ruinous state—

> with broken doors, or no doors at all, with shattered lattices, windows stopped up with boards, paper or rags, walls out of the perpendicular, pitched roofs threatening to fall... the condition of the gardens is no better than that of the building, and the whole presents a striking picture of wretchedness and decay.

The only thing they found to admire about the buildings of Srinagar was the apparent toughness of the deodar wood that had been used as beams and supports in the larger houses and mosques and bridges. In the midst of desolation this wood stood out, untouched by insects, exposure or the ravages of time. Moorcroft decided that the deodar (a type of cedar) must be the most valuable tree in all Kashmir, and their only winter expedition into the country was to collect the seeds of the deodar tree. Accompanied by two Sikh government officials for 'protection' (they knew full well that they were spies), they set off in December by boat and were paddled, poled and towed via the Manasbal lake—where they could see the ruins of Nur Jahan's garden on the north bank—into the Wular lake which, they thought, in better weather might 'afford a resemblance to some of the lakes of Westmorland or Scotland'. They stopped at the island of Lanka which the good king Zain-ul-Abidin had built. Moorcroft inspected the ruins of the mosque on the island and was gratified to be shown the engraved stone proving that the building had been erected by Zain-ul-Abidin. Then a group of ragged women emerged from their shacks to greet the foreign visitors with a song, but the travellers were not impressed. 'It was hard to say whether their squalid persons or discordant voices were more repulsive', writes Moorcroft coldly.

Beyond the Wular lake, at Sopor, they were met by their servants and horses, for at this point they were obliged to continue the journey by land. They reached the deodar forests a day later, only to find that they had missed the seeds by three months—little trees were already growing from them on the forest floor.

The boat trip had distanced them from the misery of the countryside, but travelling through it on foot and on horseback they could see clearly how derelict this once-fertile land had become. Orchards 'which must at one period have formed a forest of fruit trees' were deserted, networks of irrigation canals were choked and useless, and though the soil looked rich, very little of the land (about 1/16th in some places, and as little as 1/20th in others) was being cultivated. 'Everywhere the people are in the most abject condition; exorbitantly taxed by the Sikh government, and subjected to every kind of extortion and oppression by its officers'. No matter how far from Srinagar their route took them, the Sikhs seemed to be in evidence. At Sopor there were Sikh soldiers levying taxes, in a wretched little village they passed through there were Sikh tax-gatherers taking away no less than 9/10ths of the rice crop, and their own Sikh guards indulged in 'the usual system of violence and extortion' whenever the party stopped for long enough. The Sikhs were, almost literally, killing the goose that laid the golden egg—in their greed to squeeze the maximum revenue from Kashmir they were reducing the country to a state of such extreme poverty that Moorcroft worried about the valley becoming depopulated through starvation, disease and emigration.

Moorcroft's next excursion into the countryside was a far happier one: it was made in May the following year, 1823. The weather was warm and the breathtaking Kashmiri spring went a long way to distracting the travellers from the condition of the people. This time they followed the sight-seeing route that

Bernier had taken, to Anantnag and beyond, stopping to inspect the famous saffron fields at Pampur (not, unfortunately in bloom in May), the ruins of the ancient Hindu temples at Avantipur, and Dara Shikoh's garden at Bijbihar, where 'some of the trees were still standing, and the Chinar trees especially were of stately size and magnificent foliage. Along the centre was a line of tanks, connected by a canal, and there were also the remains of a small brick palace or lodge'.

Anantnag, which was known as Islamabad then, was a centre of weaving and embroidery second only to Srinagar. Moorcroft wrote that there were no less than 300 shawl-weavers' shops to be found there, but nonetheless 'it was as filthy a place as can well be imagined, and swarmed with beggars', so they passed through fairly quickly and went on to Jahangir's beloved garden at Verinag. This is where Bernier had been so delighted to see that some of the fish still had Nur Jahan's gold rings through their gills. Bernier's visit had taken place when the Mughals were still in residence. Moorcroft, arriving more than a century and a half later, found the garden in an advanced state of decay. Only the great octagonal pool was intact, and they walked around it on the wide stone pavement that encircles the water, and explored the arched alcoves set into the surrounding wall, and clambered around the 'mass of substantial brick-work' which was all that was left of Jahangir's palace. Moorcroft noted that most of the stone that had originally faced the building had been removed by successive generations of local people, but he did find the engraved tablet (which is still there today) commemorating the date of the building. Dotted around the garden were overgrown ruins, dilapidated watercourses, arches that had collapsed, and though their guides were eager to point out the probable site of Nur Jahan's apartments or the place where Jahangir had received his courtiers, it was impossible to form an accurate picture of how the palace must have looked in its heyday.

From the sad rubble of the Mughal palace at Verinag, the party moved on to more impressive ruins—those of the massive Hindu temple at Martand, in their magnificent site on top of a hill. 'Remarkable remains', Moorcroft called them, 'a precious specimen of ancient art . . .'.

Like Bernier, Moorcroft was anxious to visit the famous holy cave at Amarnath, particularly as he had heard that 'persons in the cave of Amarnath assert that they can hear the barking of the dogs in Tibet', but for some reason— he does not explain why—this proved impossible, so the travellers turned towards home. Shortly before reaching Srinagar, they came to the place called Pandrethan where there is a small Hindu shrine in the middle of a tank of water. Nowadays there is a bridge across the water, but there can't have been one then, for it was agreed that one of Moorcroft's party would have to swim across the murky black water of the tank and have a look at the inside of the temple. Without demurring for an instant Mr Trebeck plunged in, but, Moorcroft writes, there was no reward for such dedication. There was nothing in the shrine except a small carving of a lotus flower. Mr Trebeck must have been unobservant, or perhaps he was tired, for the roof of the shrine *is* interestingly carved. Seventy-odd years later, Marion Doughty, an exceptionally resilient and courageous traveller, visited the same site at Pandrethan and prepared herself to wade through the water to the temple, but she was no match for her predecessor, Trebeck: 'I advanced a step', she writes, and—

> gently and swiftly the black mud closed over my foot and embraced my ankle; another step forward and the dark waters, or rather the terrible clinging mud, had enveloped me to my knees. I waited for nothing further . . . I minded not that others had overcome the difficulty. I was ready to own myself vanquished, chicken-hearted, anything, if only I could succeed in escaping from the soft, hot blackness, with its terrible possibilities of writhing life and slithery inhabitants.

Moorcroft was now impatient to leave Kashmir, but there were 'vexatious delays' that held up their departure until the end of July, when at last they set out from Srinagar for Baramula on a flotilla of fourteen boats. Their carvan had been large on arrival in Kashmir, but now it was enormous. Thirty soldiers had been added to their escort and, because the group looked so well protected, a great many other assorted travellers attached themselves to it for safety. Moorcroft says: 'Our party comprehended, possibly, the greatest variety of nations that ever marched together, enrolling English, Hindustanis, Gorkas, Tibetans, Afghans, Persians, Kashmiris, Kurds and Turks, in its ranks.' Their relief at leaving Srinagar was to be short-lived. Two weeks later, when the travellers reached Kashmir's border on the road towards Muzaffarabad, an area where the local tribesmen had far more power than the governor in Srinagar, they found their way blocked by a crowd of menacing-looking armed men demanding a payment of Rs 15,000 with none-too-veiled threats of violence if they refused. Moorcroft, unintimidated, coolly offered the bandits Rs 500 'customs duty' on their goods, and not one penny more. The tribesmen refused to accept such a small sum, but Moorcroft was adamant that he would not be bullied, and in the end the only way out of this stalemate was for Moorcroft's caravan to retreat and find another route out of the country.

A month later, in September 1823, they set out from Srinagar again, this time taking the old Mughal route over the Pir Panjal pass and down to Bimber, stopping to rest in the now-dilapidated serais that Jahangir had built at convenient intervals to make the court's journey to and from Kashmir a little easier. They passed through fields of rice all ripe and ready to cut, but heard that harvesting could not begin without government permission, and permission would not be given until all the government grain stocks of the previous year had been sold at high prices. It was exactly this system which caused the terrible famine of 1877, when rain destroyed the standing rice crop before the government gave the order to harvest. Moorcroft worried once again about the country's rapidly declining population, for, driven to despair, large numbers of people were escaping from the valley, in spite of the fact that it was strictly illegal to leave without permission. Indeed, a large group of 500 refugees pleaded for Moorcroft's protection and attached themselves to his caravan. It did not take long for the Sikh escorts to grab hold of these 'half-naked and miserably emaciated' peasants and force them to work as unpaid porters, roping them together by day, and tying them up at night to prevent them escaping. At the customs post on the border the relentless Sikh officials tried to force these pathetic, bewildered Kashmiris to pay for permission to pass, and only when Moorcroft intervened were they allowed through.

Kashmir at last behind them, Moorcroft, young Trebeck and the Indian doctor Guthrie now began a long and arduous journey through Peshawar and Kabul to Bokhara. Their progress was made more difficult by the foolishly tempting size of their baggage train, for they carried all sorts of samples of Kashmiri goods, particularly shawls and textiles, not to mention the British woollen cloths that Moorcroft was hoping to start up a trade in. A rumour that they also had an enormous treasure of gold and jewels followed them across Central Asia and added to their problems. Moorcroft blamed this on an Englishman, a deserter from the army named Lyons, and he felt particularly cross about it, for he had found Lyons in 'a state of destitution' in the slums of Kashmir and had brought him out to Kabul where, instead of showing gratitude, Lyons had immediately tried to ingratiate himself with the Afghans by telling them that he could point out which of Moorcroft's chests were filled with gold. Nothing more is ever heard of Lyons. He sounds thoroughly dishonourable, but one cannot help wondering what became of him.

After eighteen months, Moorcroft's party, by now greatly reduced in both men and baggage, reached Bokhara and spent five months there before turning back towards India. The worst of their journey should have been over now, but Moorcroft had one final ambition and, like the proverbial last straw, it was to prove the undoing of everything. 'Before I quit Turkistan', he says 'I mean to penetrate into that tract which contains, probably, the best horses in Asia, but with which all intercourse has been suspended during the last five years. The experiment is full of hazard, but *le jeu vaut bien la chandelle*.'

Trebeck and Guthrie went on ahead to wait, for Moorcroft had decided to embark on this adventure alone. Why he did, and what exactly happened then is not known. It was believed for a time that Moorcroft was killed by thieves, but later investigations seemed to show that though he was indeed robbed not long after he left the others, he actually died of fever in a place called Andekhui. In any case, his servants carried his body on a camel to Trebeck and Guthrie, who were waiting at Balkh; it was only a few days since they had seen him last, alive.

The unhappy end to their story was only beginning. Very soon Guthrie sickened and died too, and Trebeck buried him beside Moorcroft. Since neither of them were Muslims, their graves had to be dug outside the city walls, and even there a mud wall was built to hide them from passers-by just in case anyone mistook them for Muslim burial places and offered up a prayer for two infidels. Young Trebeck, now alone and undoubtedly frightened in this hostile and dangerous wilderness, continued the journey towards India, but before he had travelled any distance he too became ill and died at Mazar, not far from Wazirabad. After his death the mullahs of the local mosque stole most of the expedition's few remaining possessions, but somehow Moorcroft's loyal Indian secretary managed to save his papers and made his way with them to Kabul where he handed them over to an Englishman.

Moorcroft's papers, described as 'so voluminous, so unmethodical and so discursive', were eventually published in 1841, sixteen years after his death. In the preface, their editor rather cruelly suggests that though worthy, Moorcroft's notes are 'not quite so amusing as those of some more modern voyagers', and it is true that Moorcroft keeps himself and his personal feelings out of his narrative, which makes it less entertaining. Not for him Bernier's gossipy stories nor Forster's endearing grumbles. But Moorcroft was by far the most serious and informative visitor that Kashmir had so far received from the West, and it is sad to think that during his life his efforts went so unacknowledged by the East India Company, who seem to have treated their employee as a difficult child, withholding letters of introduction to native rulers who might have helped him, scolding him for interfering with the politics of Ladakh, and stopping his salary when his journey seemed to progress too slowly. Moorcroft felt bitter about the way they treated him. The only contribution that the Company made to his last expedition, he wrote, was their 'cold permission' to go on it in the first place.

Kashmir's next known visitor from Europe, Victor Jacquemont, a Frenchman, clearly thought as little of Moorcroft's achievements as the East India Company had done—though with some excuse, for at the time of Jacquemont's trip to Kashmir in 1831 Moorcroft's *Travels* had not yet been published. Jacquemont's disparaging remarks are most intriguing for they reveal a tantalizing glimpse of another, human, side to Moorcroft, one that cannot be guessed at from his own writings. 'I live like a hermit', wrote Jacquemont in a letter home, 'and my virtue is the subject of universal admiration. Mr. Moorcroft did not set a like example of European continence here. His principal occupation was making love, and if his friends are surprised that his travels were so unproductive, they may ascribe it to this cause'. It was a little mean of Jacquemont to criticize Moorcroft in these terms, for other letters home suggest that he was not above a little dalliance himself—indeed his whole journey to India had come about as the result of a love affair. As a young medical student in Paris—a rather privileged one, whose friends were the writers Prosper Merimée and Stendhal—Jacquemont had lost his heart to an Italian opera singer, but she did not return his affections and Jacquemont was desolate. His family urged him to travel and try to forget her, so he sailed away to New York where he enjoyed a certain social success until he became involved in a dispute with a fellow Frenchman, challenged him to a duel, and was then obliged to leave America in a hurry. He took refuge with a younger brother who was a sugar planter in the West Indies, and there he learnt from his father that the prestigious botanical garden in Paris, the Jardin des Plantes, needed a naturalist to travel and collect specimens in India. One of the qualifications for the job was an ability to speak English, which Jacquemont had of course recently acquired in America—'No doubt I speak it very badly, but at any rate I speak it'.

Jacquemont returned to Paris and was selected for the post, but before leaving for India he visited London in order to establish himself as *persona grata* with the East India Company who, by that time, governed most of the subcontinent. His influential friends in France made sure that he had the best possible contacts, and he so impressed the Court of Directors of the Company that he left London with official letters of introduction to all the governors of the British possessions in India, quite unlike poor Moorcroft who was far less sophisticated and had to make his way in the world without this kind of patronage.

Jacquemont was an attractive man, especially when he put himself out to please. He was five feet ten inches tall, slender, grey-eyed, with long curling chestnut hair and a particularly beautiful voice. 'His way of making people like him was to hide none of his ideas or feelings, but to be perfectly natural', wrote Prosper Merimée. For this reason Merimée and Stendhal nicknamed him Candide, after Voltaire's ingenuous hero. On his arrival in India he charmed expatriate British society as he had charmed the Court of Directors in London, and, in particular, he made a good and useful friend of Sir William Bentinck, the Governor-General of India, and his wife. His travelling plans were vague, and Kashmir was not on his itinerary at all. But then Jacquemont had an un-looked-for stroke of luck. At that time Ranjit Singh, the Sikh ruler, employed many European officers in his army and one of these, a General Allard, hearing that a fellow Frenchman was travelling in India, wrote to Jacquemont offering to help him in any way he could. With an eye on obtaining permission to visit Kashmir, Jacquemont immediately made his way to General Allard in Lahore, and, with the additional help of letters from Sir William Bentinck, introducing him as Lord Doctor Victor Jacquemont—he was neither a Lord nor a Doctor of course—he charmed Ranjit Singh as well, and the two men had long talks together: 'He asked me a hundred thousand questions about India, the English, Europe, Bonaparte, this world in general and the other one, hell and Paradise, the soul, God, the devil, and a thousand things besides'. Ranjit Singh was obsessed by his health, and, in particular, by his impotence—a common complaint among powerful Eastern men, wrote Jacquemont, and the result of years of unbridled licence. Although Ranjit always referred to his sexual problems as 'digestive' ones, Jacquemont understood full well: 'I know what the word stomach signified in the mouth of the King at Lahore, and we discussed his malady exhaustively . . .'.

Although he describes him as a 'shameless rogue' Jacquemont obviously liked Ranjit Singh, and on his part Ranjit Singh was completely won over by the Frenchman. He is reputed to have said that he liked him better than any other European he had met. In any case, their meetings bore fruit quickly—'I am to go to Kashmir', says Jacquemont gleefully only a few days after his arrival in Lahore. 'I am to go wherever I like. The King will have me guarded everywhere'. Ranjit Singh's generosity extended beyond mere permission to visit Kashmir. He lavished Jacquemont with money and gifts, and General Allard was instructed to smooth his path to Kashmir and arrange unlimited credit for him there. Jacquemont had been plagued with money worries ever since his arrival in India, for his salary from the Jardin des Plantes did not stretch to cover even his expenses, so Ranjit Singh's open-handedness lifted a heavy burden of worry from his shoulders: 'My portable bank balance has grown much heavier owing to His Majesty's rupees; I have enough to make the journey to Kashmir and stay there for four months without making fresh inroads upon my wretched little balance in Calcutta.'

In March 1831 Jacquemont set out from Lahore for Kashmir in unaccustomed style. He now had camels and porters, foot and horse soldiers to guard him, personal servants to look after his daily needs, and a secretary—all courtesy Ranjit Singh, whose parting gifts to Jacquemont included a pair of magnificent Kashmir shawls that Jacquemont boasted would be the envy of many a Parisienne. He writes wryly: 'If Ranjit Singh feels obliged to treat his guests like this, I can understand why he is not anxious to receive visitors'.

At first Jacquemont's progress was easy. The villages they passed through contributed chickens, goats, butter, eggs and flour to feed the party, and Jacquemont could work in peace collecting plants and pressing them, examining rocks, writing notes, sketching, and looking for mineral samples. But as they marched further from Lahore the local people became less obliging and Jacquemont's time had to be spent organizing food and porters for his group. Then, not far from Kashmir, they were waylaid by a gang of wretched-looking bandits and, at knife point, Jacquemont was forced to hand over Rs 500. Only his tactful handling of the situation prevented him from being robbed of everything he carried, and killed. Jacquemont summed up the journey to Kashmir in a letter to his father: 'The Indians and Persians call Kashmir the earthly paradise. We are told that the road leading to the other one is very strait and narrow; the same is true, from all possible points of view of the one leading to Kashmir.'

Things improved on reaching the valley: Jacquemont was greeted by the Sikh governor of Kashmir with lavish gifts of food and a promise to recover his stolen money. Like Moorcroft, he was allotted Dilawar Khan's waterside garden for his camp and he found it a charming place, his only complaint being lack of privacy:

> The walls of my pavilion were merely of lace work. There was nothing to close it but shutters carved in open work with infinite art. It was open to every wind as well as to the curious eyes of the Kashmiri loungers who gathered round it in thousands in their little boats, staring at me as at a wild beast behind the bars of a cage. I had pieces of cloth stretched inside which sheltered me more or less from the wind and completely concealed me from public curiosity.

Jacquemont's plan was to base himself in Dilawar Khan's garden with occasional expeditions into the rest of the country to collect samples of the fauna and flora. Nonetheless, his work for the Jardin des Plantes was often frustrated. The two Kashmiri hunters he trained to collect specimens of wild life were treated with such contempt by their fellows, who disapproved of all the killing that had to be done, that they refused to go on with the job. At other times the heat killed the live animals he had collected, or it rotted the bodies of dead ones too quickly for them to be preserved, or the tribesmen he had recruited to help mutilated the animals in killing them so that they were useless for the collection. But what Jacquemont found far more distressing than these day-to-day professional setbacks was the misery of the Kashmiri people and the cruelty of their Sikh rulers: 'Cashmere surpasses all imaginable poverty...'. He particularly disliked the governor because the man bragged to Jacquemont that in his first year of office he had hanged 200 Kashmiris for no better reason than to frighten all the others. The people were now so cowed, he boasted, that it was only necessary to hang one or two here and there to keep the peace. 'For my part', Jacquemont says, 'if I had to govern it I should start by placing the Governor and his 300 soldiers in irons, and set them to work at making a good road'. Curiously enough, five months later, when Ranjit Singh and Jacquemont met again, Ranjit offered Jacquemont the governorship of Kashmir—an offer at which the Frenchman laughed heartily, though in some ways, he reflected later, it had been tempting for it would have been an opportunity to do so much good.

Before his detour into Sikh Kashmir, Jacquemont had travelled around British India and disliked much that he saw of the English way of life in the East. The men's drunkenness bored him: 'The English have no conversation; they sit at table for hours on end after dinner, in company with quantities of bottles which are constantly going the round.' Their extravagance shocked him: 'A young cadet who has only just landed, who does not know a word of Hindustani, has never had a service musket in his hands, and would not know how to give four men orders to march past, is sincerely convinced of the validity of his right to live like a rich man in India'. The 'repellant coldness' of the British officers chilled and puzzled him: 'I try in vain to find what bond it is that attaches their men to them. Yet discipline is admirable . . .'. And British women held no charm for him:

they may be accomplished wives and mothers, but they are nothing else. They read nothing but the *Mirror of Fashion*, a stupid periodical, principally devoted to the toilet, something like the *Journal des Modes*. They have, it is true, all the external qualities required in good society, but nothing more; and their husbands seem to be perfectly satisfied with the small talents they possess. . . . There exists scarcely any reasonable intercourse between husband and wife . . . they meet at meals, and only during the active part of that operation, for when they have done eating, the ladies are politely turned out by John Bull, who feels perfectly at his ease when they are gone. Then the bottle begins to circulate round the mahogany table. . . . Meanwhle the poor women remain in the drawing-room, amusing themselves as well they can, till the arrival of the 'lords of the creation', who sometimes keep up this circulation of the bottle so long, that when they enter the drawing-room, they find it deserted, and the lights extinguished. . . . What can you talk about to an English lady? If she attempted to join in a serious conversation, she would be immediately set down as a blue-stocking, which is a grievous affront. God preserve me from ever having an English wife.

A shrewd and witty observer of his fellow men, Jacquemont watched both the French and the English at work and play in India, and made what is surely one of the most astute definitions of the difference between their colonial attitudes: 'A Frenchman among Indians thinks to himself "I am superior". An Englishman, a thousand times richer and more powerful, thinks "I am alone".'

When Jacquemont moved from British-administered India into the Sikh territories of the Punjab and Kashmir, his criticisms of English social behaviour are replaced by a new admiration for the British system: 'One must have travelled in the Punjab to know what an immense benefit to humanity the English dominion in India is, and what miseries it spares eight millions of souls . . .'. And again:

The British Government in India, though it calls for some reforms, merits many eulogiums. Its administration is an immense blessing to the provinces subjected to it; and I have only fully appreciated it since I have been travelling in this country, which has remained independent; this is to say, it has remained the theatre of atrocious violence, and continual robbery and murder.

And in his diary Jacquemont wrote that, for their people's sake, he prayed that the desolate provinces of Kashmir and the Punjab would come under British rule as soon as possible.

An amusing side-effect of Jacquemont's disgust at the plight of Kashmir was the irritation he now felt at the poem 'Lalla Rookh'. He had brought the book with him specially to read in Kashmir, but everything he saw around him made nonsense of Moore's sugary romance, and Jacquemont became infuriated by both the poem and its author. Thomas Moore, he writes to his family at home, was 'a perfumer and a liar to boot'. Moore's poem had given the impression that Kashmir was a land of smouldering dark-eyed beauties. Jacquemont explodes: 'Be it known to you that I have never seen such hideous witches as in Kashmir . . . Were I manager of a theatre, or of a strolling company performing *Macbeth*, I should have but little trouble finding witches: for I meet plenty every day'. Later Jacquemont simmers down and qualifies his remarks by explaining that the lack of pretty women in Kashmir was undoubtedly because all little girls showing promise of good looks were sold at the age of eight and exported to the Punjab and India: 'They are sold by their parents for 20 to 300 francs, the average price being 50 or 60 francs'. The girls became slaves, but Jacquemont believed they were usually fairly well treated, and their position was 'hardly worse than those of their mistresses in the harem'.

The summer of 1831 was unusually hot, and in August Jacquemont moved from his garden pavilion to Char Chenar, the Isle of the Plane Trees, to try to catch a breeze. The little island had been a favourite Mughal picnic ground: four plane trees had been planted, one at each corner of the island, and a pavilion built in their shade. Jacquemont found only two of the trees left, but the now-derelict pavilion in which he camped gave shade enough—he describes it as 'one great hall open to all the winds when they are pleased to blow, with a ceiling supported on columns in a fantastic style, carried off from some ancient pagoda'. From his new island home, Jacquemont could look across the lake and see the sun glinting on the minaret of the mosque at Hazratbal, where a hair from the Prophet's beard was preserved. (The mosque has been rebuilt and the hair is still there today.)

Not far from Hazratbal the huge plane trees of Akbar's garden, Nasim Bagh, made a dark patch at the foot of the hills, and across the water Jacquemont could make out the avenue of poplars at the Shalimar garden. The heat made it impossible to work, so he swam in the warm limpid water of the Dal lake, accidentally burnt his body a 'vivid crimson', and celebrated his thirtieth birthday in a reflective mood. 'I am looking well', he reassures his family, 'but I am terribly thin'. In Delhi he had bought a mirror, but he apparently rationed himself to looking in it only once a month.

Jacquemont was touchingly fond of his father and brothers at home in France and every letter home is full of love and tenderness towards them. He yearns for their news, and chafes at the length of time it takes, ten months or more, for letters to arrive. He watches for the postman—in his case runners sent up from Lahore by General Allard—as eagerly as any schoolgirl. From time to time his family must have expressed concern about his health and the dangers of the life he led, for Jacquemont gently teases his father about the old man's fears: 'I think a man must be rather foolish to allow himself to die at 30; and I have the vanity to believe that I shall not commit such a piece of folly for a long time to come'. It is pathetic to read his brave protestation while knowing in advance that the end of his story was a tragic one—for Jacquemont never saw again the family he loved so much.

In September 1831 Jacquemont left Kashmir to make his unhurried way to Bombay, where he planned to ship the collections he had made to the Jardin des Plantes. His caravan was a very much weightier one than it had been on his arrival six months before. The porters were loaded with cases, barrels and bundles of animal, plant and rock specimens, as well as two huge trunks packed with Kashmir shawls and other souvenirs, destined for Jacquemont's family and friends.

They marched first to Amritsar where the Frenchman met up again with his old benefactor Ranjit Singh, and then on through Delhi, where he was irritated to find that the expensive barrels he had had made and filled with turpentine to preserve his fish specimens were leaking and needed repair. Next, to Rajasthan, which he found delightful, relishing its picturesque palaces and cities, particularly those of Jaipur and Ajmer. But Rajasthan was the last happy stopping-place for Jacquemont. As he travelled further south he found the heat oppressive and exhausting, and he thought with longing of the clear air of Kashmir and its cool shade and dappled waters. Worse still, he found that as his journey continued the British he met were less and less to his liking—they were coarser, he wrote, and not as well-bred as his friends in the north.

In Poona he became violently ill with dysentery and, though he survived the attack, found it difficult to throw off its effects. Then a cholera epidemic broke out in Bombay and Jacquemont was advised to delay his arrival there for five months. He became depressed, time hung wearily on his hands, and the British he met showed no interest in India: they knew nothing about the country and cared less, they spoke none of its languages, had not travelled and had no curiosity in Jacquemont's experiences. 'Ah what stupid creatures the people at Poona are, my dear fellow', he writes to a friend—

> They go out riding and driving, breakfast, dine, dress, shave and undress, or meet on committees for settling the affairs of a public library where I have never seen anybody but myself... and that is their whole life: The stupid creatures! The idiots! The judge is a perfect idiot; the magistrate a rabid hunter, etc. etc. The only sensible man is the general.

At a dinner party he met a fellow naturalist—Baron Hügel ('What German is not a baron, outside his own country?', asks Jacquemont sourly). The two should have had a great deal in common, but Jacquemont wrote that though he plied Hügel with questions that any naturalist should have been able to discuss, the man did not appear to know what he was talking about. Baron Hügel was clearly wounded by these comments, for in his own volume of travels, published a decade later, he criticizes the decision to publish Jacquemont's private letters: 'they have done immense injury to his memory in India. They who assisted him with advice and their good services, now stigmatize him as an ungrateful slanderer, and many, who really liked and esteemed him, feel deeply hurt and disappointed.'

At long last Jacquemont was able to move on to Bombay, only to become desperately ill the very day after his arrival there. The doctors diagnosed an abscess on the liver, and Jacquemont lay ill and in pain for a month before he could summon the energy to write a letter home to his brother. It was to be the last letter he wrote.

> The most cruel thing, dear Porphyre, when we think of those we love dying in distant lands, is the thought of the solitude and abandonment in which they may have spent the last hours of their life. Well, my dear, you must find some consolation in my assurance that, since arriving here, I have never ceased to have the most affectionate and touching attentions heaped upon me by a number of kind and friendly men.... My end, if it is that which is approaching, is easy and tranquil. If you were here, sitting on the edge of my bed with Father and Frederic, my spirit would be in anguish and I should not be able to watch the coming of death with such resignation and serenity.. . . Adieu! Oh, how your poor Victor loves you! Adieu for the last time!

Five days later, at the age of only thirty-one, Victor Jacquemont died and was buried in Bombay with military honours—for he had been made a member of the French Legion of Honour during his travels in India. His affairs were taken in hand by a Mr James Nichol, an English merchant and one of those 'kind and friendly men' Jacquemont mentioned in his letter. Mr Nichol arranged for the packing of all Jacquemont's carefully collected natural-history specimens, and eleven large cases and a barrel were shipped back to France, together with a box of Jacquemont's notes and papers, and a tactful letter to his family breaking the news of his death.

In the year following Jacquemont's visit, a less attractive character arrived in Kashmir from Europe: thirty-seven-year-old Joseph Wolff, the son of a Bavarian rabbi, who had first studied for the Roman Catholic priesthood in Rome and then became a Protestant and settled in England. Actually, 'settled' is hardly the word to use for the eccentric Mr Wolff, for he had dedicated his life to the conversion of Jews around the world and—first as a one-man effort, and then under the auspices of the London Society for Promoting Christianity among the Jews—he travelled to America, Abyssinia, Armenia, Turkey, Persia, Afghanistan, and India, preaching, praying and arguing as he went. Wolff was not a humble man and his memoirs, pompously written in the third person, are really a long monotonous blow on his own trumpet. They consist of obscure religious arguments in which he did well, neat points that he scored off rabbis or mullahs, and his prodigious feats of preaching Once in Calcutta he apparently talked for no less than twelve hours a day for six whole days, an appalling thought. The account of his visit to Kashmir adds little to knowledge of the country at that time.

Sher Singh, Ranjit Singh's son, was Governor of Kashmir at the time of Wolff's visit in 1832, and was as callous as most of the other Sikh and Afghan governors. On the path into Kashmir Wolff met women and children 'howling and weeping'. When he asked why they were leaving Kashmir they replied: 'On account of the tyranny of the rulers . . . we are bought and sold like pieces of bread.'

Sher Singh welcomed Wolff in the Shergarhi palace in a splendid room with glass chandeliers hanging from the ceiling and Kashmir shawls spread over the floor: 'There were all the grandees of Kashgar, Kikan, Khotan, Ladack, Lassa, present: also some Chinese, Persian Moollahs, Pundits and Brahmins. . . .' Although he was shocked at the cruelty practised by the Sikhs against the Kashmiris, Wolff managed to spend his short visit 'both pleasantly and usefully' in religious discussions.

Perhaps for all his self-importance Wolff did have a sense of humour buried somewhere. Certainly he chuckled over the Meerut newspaper account of his arrival at Government House in Simla which described him as 'Joseph Wolff the perverted Jew' instead of the 'converted Jew', and he tells a nice story against himself. One day when preaching in Kashmir he found himself surrounded by beggars, all assuring him in flattering tones that the fame of his name had gone far before him. If that was so, Wolff said, could they please tell him his name—but of course none of them knew it. Wolff must also have had a soft heart, for when he left Kashmir he begged Sher Singh to relax the rule forbidding Kashmiris to leave the valley, and allow a party of poor people to accompany him out of the country. Sher Singh agreed,

> and so it came to pass that hundreds of shawl-weavers, with their wives and children, joined Wolff on his journey back. . . . They came close to his palanquin, and the police tried to drive them back with their sticks: but Wolff suddenly jumped out of the palanquin with a stick in his hand, and said 'Do you dare to disturb the companions of the great Englishman?'

At the end of his travelling days Wolff settled down as vicar of Ile Brewers in the West Country of England. One can only hope, for the sake of his parishioners, that he had lost his fondness for the sound of his own voice.

In November 1835 a positive crowd of European travellers descended upon the valley of Kashmir. The Austrian naturalist Baron Hügel, whom Jacquemont had met in Bombay, arrived from India via Jammu, Kashmir's neighbouring state; Dr Henderson, an employee of the East India Company, trudged in from Ladakh where he had been held prisoner for three months; and Mr Godfrey Vigne, a gentleman who worked for no one but travelled simply to satisfy his own curiosity, trekked in from Skardu. It was an extraordinary coincidence that brought the three men together on the evening of 18 November in Dilawar Khan's garden which, like Moorcroft and Jacquemont before them, the travellers had been allocated by the authorities.

Godfrey Vigne had been the first to arrive in Kashmir. A thirty-five-year-old Englishman, he had thrown up his career as a barrister in London some years before in order to travel and write and draw, and he illustrated his own books with great skill. His first journey had been through America and Canada, but in 1833 he had turned his attention to the East and travelled to India, from where he was to spend the next seven years venturing into the wildest of places in the north-west of the country and beyond. Vigne had been out shooting wild duck on the Dal lake the day Baron Hügel arrived in Srinagar, and he returned to the garden camp to find not only the Baron, whom he had been expecting, but also an extraordinary, battered-looking figure wearing the costume of Ladakh. This was the eccentric Dr Henderson who, earlier in the afternoon, had startled Baron Hügel.

Hügel had arrived in the garden, settled himself in, and was writing up his diary while he waited for Vigne to return from the duck shoot, when a servant interrupted him with the news that 'a man, most wretchedly clad, without doubt some Englishman', desired to speak to him. The Baron, a refined person, was truly shocked by the creature who was then ushered in.

> His long red and white face, prominent nose and eyes, with matted red beard, constituted his chief personal peculiarities: his filthy tattered garments were partially arranged according to the Tibetan costume. In the strongest Scottish accent, he begged my pardon, and said he expected to find Mr Vigne there. I exclaimed involuntarily 'Who on earth are you?' To which he replied, 'you surely must have heard of Dr Henderson?'

Very few people in northern India at that time had *not* heard of Henderson, for the doctor had an insatiable craving for adventure and was forever taking leave of absence from his post with the East India Company and setting off on hazardous journeys into the unknown. These lone expeditions did not make him an easy employee, and inevitably they landed Dr Henderson in scrapes from which he was often lucky to escape with his life. He seems to have been an enthusiastic student of disguise rather than a master, for, no matter how he dressed himself up for these journeys, his true identity was discovered again and again. Even Hügel's servant had described him as 'a man, most wretchedly clad, without doubt some *Englishman*', and the Baron wrote that whenever Dr Henderson went on his travels the British authorities would receive information from far-flung mountain regions that 'an Englishman in disguise' had been seen. In Kashmir there was no reason at all for Henderson to disguise himself, but nonetheless he swapped his Tibetan costume for a Kashmiri one—'If not very magnificent, it was at least clean', remarked Hügel fastidiously. This turned out to be an error: the local people *were* taken in by his appearance for once, and while they would talk readily and unguardedly to Hügel—an obvious foreigner in his European clothes—to Henderson, now looking like a fellow-Kashmiri who might at any moment demand money from them under some pretext or another, they revealed nothing. This annoyed Henderson greatly and gave Baron Hügel a certain satisfaction: 'It is not always to our advantage to assume the native dress', he comments smugly.

But on that Wednesday evening in November when the three men met so unexpectedly, there were no differences of opinion—they were clearly overjoyed to be together. Vigne held Henderson in great respect and declared that admirers of adventurous travel could not fail to include him in the 'highest rank', but one has the feeling that Hügel, though also impressed by Henderson's fortitude and endurance, was faintly appalled by his uncouth habits and rough appearance. Perhaps seeing himself as the aristocrat of the party, or perhaps simply because he was better supplied, Hügel appointed himself host to the other two travellers during their stay in Srinagar and provided their evening meals. One night the party dined off hare soup, fresh salmon, roasted partridges, and a ham from the wild boar of the Himalayas. Hügel, with quiet pride, wrote that the meal was 'better than usually falls to the lot of travellers in Kashmir'.

A handsome man of forty with dashing upswept moustaches, Hügel had led an extraordinarily varied and adventurous life. After studying law at Heidelberg, he had enlisted in the army and travelled to France and Italy, Sweden and Russia. In 1824 he had retired from military life to his home in Vienna to concentrate on his real interest—horticulture and natural sciences. Then, like Jacquemont, a broken heart drove him to travel abroad. In Hügel's case the woman he loved was snatched from under his nose by Prince Metternich, but Hügel does not seem to have held this against him. Long afterwards, during the 1848 political upheavals in Europe, when rioting mobs stormed through Vienna and Metternich's government fell, it ws Hügel who smuggled Metternich and his wife out of the city in his carriage and took them to England.

Hügel's travels took him to India in 1832 and then, in the following two years, on to Indonesia, Australia and New Zealand, and back to India again—when he made his journey to Kashmir. As he progressed through all these countries he collected assiduously, so that by the time he returned to Europe he brought with him no less that 32,000 scientific specimens, including 12,000 plants, and 3000 different varieties of seeds, as well as musical instruments, manuscripts, coins, weapons, and anything else he considered of interest. These he presented to the Austrian state museums while he went home to Vienna to become the country's best-known and most influential gardener. Indeed the parks and gardens of Europe owe a great debt to Hügel for the many new plant species he brought home from his travels and raised in his nurseries.

On Hügel's journey into Kashmir, despite all his vast experience, things had a way of going wrong for him that seems, at times, positively comic. Hügel was determined to travel to the valley in style—'so far as physical comforts could be provided, I resolved to want for nothing', so the Baron took 'meats hermetically sealed in tin boxes', wines and other drinks, preserved fruits and sweetmeats and a retinue of no less than thirty-seven servants, including a special man to look after his hookah pipe, a tailor, a cook and two assistant cooks, two huntsmen (for catching specimens of game), two gardeners (to collect plants and seeds), two butterfly catchers, twelve bearers to carry his sedan chair, and a herald and two messengers who wore breast plates with Hügel's name engraved on them in Hindustani and Persian. But finding enough men willing to haul all their accompanying bag and baggage through the mountains to Kashmir was quite another matter, and to Hügel's great irritation his porters sulked, went on strike, dropped his camp bed on purpose and broke it, and even, occasionally, ran away. Hügel had to become a human sheepdog, keeping the caravan moving by snapping and worrying behind them. Once, he tells us, he had to thrash five of the porters with a bamboo cane to get them started. In the mean time the cook had to be dismissed for fighting with the other staff, but not before he had bitten one of the messengers quite badly on the hand.... And all the while Hügel was involved in various other endearing misadventures. One night he sent his huntsman off after a fox-like creature he had spotted in the bushes. The man took so long to return that Hügel decided to go after the fox himself, so he crept off into the undergrowth, and was then shot by his own hunter. Luckily no harm was done. On another evening, while passing through a small village, Hügel unthinkingly lifted his rifle and, for sport, shot down a vampire bat. He was then very nearly lynched by a howling mob of furious inhabitants who, it turned out, had worshipped the bat as sacred. In another village Hügel discovered that there was a species of bird that the locals believed could not be killed. Determined to prove to them that this was a silly superstition, both he and his hunter tried to shoot down some of the birds but, to the delight of the villagers, first their gun jammed and then they missed.

But the incident most typical of Baron Hügel's kind of luck happened at Pampur in Kashmir, when he took a hammer and swung at a rock that seemed likely to conceal some interesting fossils. The hammer broke instead of the rock, and Hügel was forced to wait for hours while a minion was despatched to a nearby town for another one.

Hügel bore with all these pinpricks as part and parcel of the traveller's lot. What he was not expecting at all, though, was the crushing disappointment of his arrival in Kashmir. It had been his lifelong ambition to visit the place, to see whether the country lived up to its romantic reputation, or, as he himself put it, whether Kashmir would 'bear the uplifting of the veil which has so gracefully and immemorially hung over her'. His anguish was acute when he found that, for him at least, it did not: 'All that I saw during my first day's stay in Kashmir, was the ruins of what once had been palaces, old dilapidated houses, streets of unexampled filthiness; a population strictly corresponding with them... such were my first impressions of this long dreamt of fairy land'. The piercing cold in Dilawar Khan's garden pavilion kept the poor Baron awake most of his first night, and he had plenty of time to reflect on his miserable feelings of anti-climax: 'Reluctantly, I acknowledge that my arrival in Kashmir has not afforded me any satisfaction...'. But when, next day, the Baron lifted his eyes from the squalid city to the grandeur of the mountains around, the natural beauty of the place restored his spirits enough for him to start making sightseeing plans with Vigne and Henderson. He decided that four days was quite long enough to take in the city of Srinagar and its surroundings, and that after that he would make off into the countryside.

On their first day together, Vigne, Hügel and Henderson pottered around the Mughal gardens in the Dal lake. At Nishat Bagh, Hügel was struck by the prettiness of a series of pavilions built on arches over the canals (sadly, these—as explained in Chapter One—no longer exist). He found little else to admire except the huge plane trees, for the garden was 'in other respects, almost a perfect wilderness'. At the Shalimar garden they found the black marble pavilion built by Shah Jahan in reasonably good repair as both Afghan and Sikh governors liked to use it as a place for picnics and receptions, but the rest of the garden was unkempt, and there was an untidy hamlet of huts encroaching on the land.

The three men then moved on to Char Chenar, the tiny island where poor doomed Victor Jacquemont had camped to try and escape the burning summer heat four and a half years before. Hügel made a note about the dilapidated Mughal pavilion in which Jacquemont had lived—'a single open hall, with a little tower, commanding a fine prospect of the lake', but the travellers were more impressed, it seems, by an ancient water-wheel they found under one of the chenar trees. Towards the end of the day the three progressed to Nasim Bagh, Akbar's great shady garden of chenar trees. Vigne complained that according to local custom they should have arranged the visits the other way about—the tradition was to visit Nasim Bagh in the morning when the huge trees offered protection from the heat, and Nishat in the cool of the evening.

Akbar was supposed to have planted more than 1000 chenar trees at Nasim Bagh, and Hügel marvelled: 'They are still in fine preservation, although planted more than 200 years ago, forming beautiful walks, whose refreshing shade in summer must be delicious'. But Vigne wrote sadly: 'a great number of these fine trees have been destroyed by the Sikhs. The governor, Mehan Singh, cut down some in the Shalimar, and sold them'. Even the hated Afghans had, apparently, tried to preserve the trees, for Vigne adds: 'In the time of the Pathans no man could cut down a Chinar under a penalty of 500 rupees, even on his own ground'.

Next day the three climbed to the top of the Takht-i-Suleiman (or Shankaracharya) hill which dominates Srinagar. A picnic meal was carried up to them and they passed the whole day there with Vigne making sketches for a view of the whole valley, which was to be shown at the Panorama in London. Their third day was taken up with a compulsory audience with the Sikh governor, Mehan Singh. The state boat was sent to collect and take them to the governor's residence, the Shergarhi palace, and they were received on an open terrace by the governor whom Hügel disliked intensely on sight and described as a dissolute man with 'thick lips and but half opened eyes'. His bodyguards were lavishly and colourfully dressed, but what caught Hügel's eye was a group of unhappy young men seated on the floor. These were hostages from the ruling families of the more remote mountain areas that Ranjit Singh now controlled—the tribes were far less likely to give trouble knowing their best young men were prisoners in Kashmir. Hügel saw that the young princes were burning with resentment at being made to sit at the foreigners' feet, and that the governor was revelling in their humiliation, taking pains to introduce each one in turn to the three Europeans. There were no other foreigners at the court such as Wolff had described—no Persians, no Chinese, and no Kashmiris of any consequence, for 'the country is so completely subjugated, that the natives, except a few traders in shawls, are nothing better than so many beggars'.

After their unpleasant audience with the governor, Hügel, Vigne and Henderson decided to explore the old town of Srinagar by boat, stopping now and again to examine the bridges and mosques they passed. To his great irritation Vigne suddenly realized that he had left his sketchbook behind at Dilawar Khan's garden, and a servant was sent back to fetch it. While Vigne waited in his boat for the sketchbook, Baron Hügel and Dr Henderson went on ahead to inspect the Pathar Masjid, the pretty but sadly neglected mosque built by Jahangir's wife Nur Jahan. Now a mishap worthy of Baron Hügel himself took place—Vigne was so eager to catch up with the others that, as soon as his sketchbook arrived and his boat was being rowed towards them, he jumped up with such enthusiasm that he toppled into the freezing river. 'A cry of terror burst from the crowd of assembled natives on the shore', writes Hügel dramatically, but Vigne was an excellent swimmer and astonished the spectators by swimming round and round to warm himself before getting out of the river.

Wet through and cold, Vigne had to be ferried back to Dilawar Khan's garden to change, while Hügel and Henderson continued their sightseeing on foot, wending their way through the dirty narrow back streets towards the Hari Parbat hill on the outskirts of Srinagar. On the way they noted several large buildings that had once been the homes of nobles at the Mughal court, but, apart from those which had been turned into shawl factories, they were 'deserted, lonely ruins'. At the foot of Hari Parbat hill their walk took them through the ruins of Akbar's city, Nagar-Nagar—long since deserted and tumbledown. Hügel says:

> Blocks of stone and large columns, brought from the more ancient temples of Kashmir, lie in desolate grandeur around. A beautiful mosque, built by Achan Mullah Shah, deserves to be mentioned,

particularly on account of the finely wrought black marble and stone lavished upon it. The gates are made of one single stone, and polished like a mirror, but the wanton love of destruction during latter years has torn some out of their places, and others lie perishing on the earth.

(This is the charming mosque, built by Jahangir's son Dara Shikoh for his spiritual mentor Akhund Mullah Shah, which is mentioned in Chapter One.)

Having walked the whole way to the top of Hari Parbat hill, Hügel and Henderson were not best pleased to be refused entry into the fort there by military guards—the same thing, incidentally, that happens to modern tourists who do not have a written pass from the Kashmir Tourist Office. On the way down again they were re-joined by Vigne, now changed and dry and no worse for his icy dip, and together the three men decided to visit a shawl factory. But this was a distressing experience, for though the sixteen craftsmen they saw were weaving a pair of exquisite shawls that would sell in Kashmir for Rs 3000, they worked in appalling conditions, huddled together in what Hügel described as 'one of the most wretched abodes imaginable'.

By now Hügel had had enough of Srinagar and its environs and he and Henderson set off by boat towards the riverside town of Anantnag. Their route followed Moorcroft's, taking in Pampur, the ruined temples of Avantipur and Martand, and a whole series of ruined Mughal gardens, including of course Verinag with its great octagonal tank and Achabal with its dramatic gushing spring. The pavilions in these gardens were in ruins, which made it hard for Hügel and his party—'fallen roofs so blocked up the interior as to leave little room for a wanderer's observation'—but they were possibly more interesting in that state than the tidied-up versions we see today. The weather was extremely cold and the trip most uncomfortable, and Hügel and Henderson were thankful to be back in Srinagar with Vigne five days later.

Time was running out for them now: all three men were due to leave the valley within the next few days. But Hügel felt that their meeting had been so extraordinary that they should not part without commemorating it in some way—perhaps, he suggested, by putting up a suitably engraved stone. After some discussion it was decided that a monument to themselves would look too egocentric, and that they should include on their memorial the names of past travellers to Kashmir, so Hügel drafted the words:

THREE TRAVELLERS:
BARON CARL VON HÜGEL, FROM JAMMU
JOHN HENDERSON, FROM LADAK
GODFREY THOMAS VIGNE, FROM ISKARDO
WHO MET IN SIRINAGUR ON THE 18TH NOVEMBER 1835
HAVE CAUSED THE NAMES OF THOSE EUROPEAN TRAVELLERS
WHO HAD PREVIOUSLY VISITED
THE VALE OF KASHMIR, TO BE HEREUNDER ENGRAVED—
BERNIER, 1665
FORSTER, 1786
MOORCROFT, TREBECK AND GUTHRIE, 1823
JACQUEMONT, 1831
WOLFF, 1832
OF THESE, THREE ONLY LIVED TO RETURN TO THEIR NATIVE COUNTRY.

His list of 'previous travellers' did not include the Portuguese Jesuits who visited Kashmir—'I need not remark', writes Hügel aggressively, 'that in the list I have included no Catholic missionaries.'

Vigne, being the most artistic member of the group, was put in charge of the design and lettering of the stone. As he was printing out the words on paper he complained that there would be no room for the inscription unless he made either their names or the names of the previous travellers smaller. 'Oh damn, the previous travellers: get in our own names as large as you can', replied one of the others—Vigne is too tactful to say which—and laughingly, they all agreed.

Now the only outstanding problem was acquiring a suitable stone for engraving. Hügel, showing a sudden and total lack of sensitivity, particularly since he had so recently bemoaned its sorry state of repair, suggested that they should use one of the black marble doors of Akhund Mullah Shah's pretty mosque at Nagar-Nagar. The others seem to have accepted this without a murmur, and Vigne went over that evening to collect the door.

But he forgot to take any ropes or poles with him, and the marble door was too heavy to shift. Next day Vigne's servant, Mitchell, was dispatched to Nagar-Nagar with the appropriate tackle, but he too came back empty-handed, saying that a huge crowd had assembled to prevent the door being removed, and a guard had threatened to shoot if Mitchell didn't leave immediately. Hügel, who had very little time for Mitchell—'a half-caste and a confirmed drunkard'—did not believe a word of this excuse, but two days later his own servants warned him that the city was up in arms at the idea of the door of the mosque being taken, and that 'nothing but our robbery was talked of in the bazaar'. The three hastily decided to look elsewhere for their stone, and a suitable slab was found in the Shalimar garden—one cannot help wondering what *that* belonged to. A sculptor was hired to carve the inscription, but they were not allowed to erect the stone on the Char Chenar island, the site they had chosen for it without Ranjit Singh's special permission, so the stone was given into safe hands with instructions that it should be put up the moment the permission came through.

On 2 December Dr Henderson bade goodbye to the others and marched off, bound for Attock and the southern side of the Hindu Kush—wild, difficult and dangerous country. He died of fever a little more than a year later.

The following morning Vigne and Hügel, who had decided to join forces and travel together to Lahore, set out on their less arduous journey. Before they left, Vigne made a sketch of Dilawar Khan's garden, complete 'with every inhabitant at present in it, servants, horses, dogs, goats, and poultry', and presented it to a delighted Hügel as a souvenir of their stay.

Less than two years later, in 1837, Vigne passed through Kashmir again on his way back to Ladakh and found that their stone had *not* been erected. He patiently re-applied for permission to put it up, and when that came through, he himself supervised the placing of the memorial on Char Chenar island. Vigne was glad to see that the stone was still in place on his final visit to the valley in 1839, but he feared for its future and thought it extremely likely that 'the first of my countrymen who looks for it, will find that it is gone. Should such be the case, I cannot lose this opportunity of requesting him to replace it; and if he has the *esprit de corps* of a traveller, there is no occasion to give him any reasons.'

No one seems to know how long the memorial survived, but it is certainly nowhere to be seen today. Some people believe that in a fit of anti-British fever around the time of Indian Independence it was hacked off the wall and thrown into the Dal lake. One day, maybe, it will be found somewhere—or perhaps some Romantic with the *esprit de corps* of those marvellous Victorian travellers will take up Vigne's invitation to replace it.

CHAPTER THREE

Resort of the Raj

I must go 'home' now to England. Home? What mockery! If home be 'where the heart is'... mine I have left in the 'pathless deserts', and among the stupendous glaciers, far, far, away.

—Mrs Hervey, *Adventures of a Lady in Tartary, Thibet, China and Kashmir* (1853)

CHAPTER THREE

Resort of the Raj

The most touching reminder that Kashmir was once a playground of the British in India is the corner of it that is forever England—the old Christian graveyard in the area of Srinagar known as Sheikh Bagh. Apart from its extraordinary Englishness the cemetery does not look particularly fascinating; in fact, the older graves are not even in sight, for they lie beyond a lych gate in another secluded section of the cemetery—and even there, the oldest one, that of a British Colonel of the 9th Lancers, dates from only 1850, which is positively recent compared to many another British grave in India. But it is not because of its age that the cemetery is interesting: it is because its tomb inscriptions give us such a very poignant picture of who the living once were, and what they were doing in this Himalayan valley so far from home.

All the eager sportsmen who went to Kashmir on leave from India determined to bag some impressive trophies are represented in the cemetery by Lieutenant Robert Lloyd Edwards of the 4th Battalion Rifle Brigade, who met his death 'falling over a precipice while out shooting at Bandipoora, Kashmir, on 19th October, 1877', and by Colonel Sydney Drummond Turnbull of the 15th Bengal Lancers, who died in Kashmir in February 1911 'from wounds received in an encounter with a leopard'.

All the professionals—the engineers, surveyors, foresters, soldiers, administrators and accountants—who served the maharaja's government in one way or another, and whose wives did their best to make their 'quarters' feel like home—bravely gave birth and tried to raise their children against the odds—are represented by the many graves of still-born babies and young children killed by typhoid or cholera, particularly by the pathetic tomb of the Srinagar postmaster's children on which the inscription reads: 'In loving memory of our dear little pets Lena Myrtle aged 2 years 4 months and Eustace Livingstone aged 9 months. The darling children of Arthur and Margaret Appelby, who died 30th May 1893'. All the missionaries—doctors and teachers—who tried to introduce their Christian God to the Kashmiris through hard work and humanity are represented by the grave of Fanny Butler, the first woman doctor in Kashmir, who died of fever only a year after her arrival in Srinagar. All the loneliness and precariousness of lives lived far from England are summed up on the tombstone of twenty-one-year-old Frances Ramsay, whose sparse inscription just says: 'Held in high esteem by Brother Fred'. Even today's Western tourists and holidaymakers are represented, for this is a living cemetery (if one can say such a thing). It is still used as the burial place for Christian foreigners who die in Kashmir.

The cemetery bears witness to the extent of the British involvement in Kashmir, and yet the British role was never quite the all-powerful one that some of the earlier travellers had predicted. In 1835 Godfrey Vigne had confidently assumed that in the very near future the British would seize Kashmir for themselves, that their flag would fly from the ramparts of the Hari Parbat fort, and that British skills would develop the valley's agriculture, industry and mineral resources until it became 'the focus of Asiatic civilization: a miniature England in the heart of Asia'. Mrs Hervey, an independent lady whom we will meet later in this chapter, visited Kashmir in 1853 and included many descriptions of it in her book of travels, for she felt that these would come in useful when the British took over, as she fully expected them to do as soon as the first maharaja died. But, rather surprisingly, the British never reneged on their treaty with the maharaja, or his descendants, and the Union Jack never did fly from Hari Parbat hill. The prayers that Victor Jacquemont, the French traveller, had offered up, imploring the Lord to save Kashmir by making the British take it over, were never answered, but he would not have been altogether disappointed with what did happen. For although Kashmir remained a princely state, and was never fully absorbed into

'British India', numbers of Englishmen went there to advise and assist the various maharajas, and they were usually of the type that Jacquemont most admired. Jacquemont firmly believed that the British colonies owed their strength and prosperity to what he called the 'activity, industry, order—in fact, to the social superiority of the majority of the men who founded them and their descendants...', and he despaired of his own country, France, which, he wrote, peopled their colonies with 'fencing-masters, dancing-girls, hairdressers and milliners'.

In 1906 the year-round British community in Srinagar numbered about sixty or seventy souls altogether, but in the summer months, when people came up on leave from India, there were of course many more. To iron out the difficulties threatened by this annual influx of foreigners, the Kashmiri government set up what must have been one of the earliest tourist offices, the Motamid Darbar, at which all visitors were obliged to register themselves. The Motamid Darbar helped new arrivals to find accommodation and staff, set the rates of pay for coolies and boatmen, arbitrated in disputes, and published annually a set of *Rules For Visitors to Kashmir*. These stated, among other things, that the road into Kashmir through the Banihal pass was reserved for the exclusive use of the maharaja, that visitors were responsible for their servants' debts, that uniform must be worn at state banquets, that fishing was prohibited in all sacred tanks, and that no visitors might occupy houses in the town of Srinagar.

The first of the maharajas, Gulab Singh, had set aside a couple of shady neighbouring gardens—Munshi Bagh and Sheikh Bagh—along the river outside the city for the use of foreigners, and these quickly developed into a busy European Quarter. Before the nineteenth century was out, Munshi Bagh (which is the area around the Bund) boasted a cluster of guest-houses (known as the Barracks), bungalows for missionaries working in the Christian Missionary Society hospital nearby, a club and a public library—which had the added attraction of a vegetable garden run by the Library Committee. (According to a Mrs Burrows, writing in 1894, 'a very useful basket of vegetables can be purchased there for 4 annas, its composition varying with the season of the year'.) All Saint's Church and the parsonage went up in 1896—(apparently *the* social event that summer was the Building Fund Fête)—by which time the Residency had already been re-built with handsome carved pillars, wainscots and ceilings. The original Residency building in Munshi Bagh had been severely damaged in an earthquake in 1885, when it was only a few months old.

Today the Residency houses the Kashmir Handicrafts Emporium, the Club is still the Club, but the old post office, though it still stands, has been replaced by an ugly modern one. Grindlay's Bank, in black and white 'Stockbroker Tudor,' is still there, looking as though it would be more at home in Surrey. The Mission Hospital, run-down and sad now, has become the T. B. Sanatorium, and only the tower of All Saint's Church is left—the body of the building was burnt down in riots some years ago.

After a fearful flood in 1893, the river bank was strengthened to protect Munshi Bagh, and the new embankment or 'bund', as it is still known today, became a favourite walk for the European ladies and gentlemen. Mrs Burrows' first sight of Munshi Bagh was of 'white tents peeping out from the trees . . . with nice clean English faces walking about amongst them . . .'.

Shops, agencies and banks soon sprang up to service the ever-increasing number of Europeans, and behind Munshi Bagh a polo ground, cricket pitch, tennis courts and a hockey field encroached on the 'maidan'—the 'plain' or public space frequently found in Indian cities. Beyond them, Nedou's Hotel was opened in 1900.

The other garden, Sheikh Bagh, was an enclosed orchard a little further downstream from Munshi Bagh. There things were quieter, for, apart from a guest-house which the maharaja lent to only the most distinguished of his visitors, the permanent occupants were the Christian dead in the cemetery, and some missonaries—including Cecil Tyndale-Biscoe and his family, whose very English-looking home, Holton Cottage, still stands near the cemetery enclosure.

Outside Srinagar, visitors could moor their boats more or

less where they wished, but in the city it was the custom for women and families to base themselves in Munshi Bagh or Sheikh Bagh—where the houseboats soon lined both sides of the river—while bachelors were allotted their own camping ground called Chenar Bagh, on the other side of the playing fields, behind Nedou's Hotel. Miss Parbury, an English visitor in 1902, was thoroughly piqued at this segregation of the sexes. From her mooring in the Munshi Bagh she wrote crossly,

> There is another more beautiful and shady place called the 'Chenar Bagh', or 'Chenar Gardens' which, the guide-book tells us, is reserved exclusively for a bachelors' camping-ground, and, however much tempted by its beauty, ladies are not expected to tie up their boats there, which shows that the bachelors are both selfish and unsociable.

Miss Parbury's thoughts seem to have lingered on these bachelors, and one suspects she imagined all kinds of wild goings-on in the Chenar Bagh. Returning from an excursion one day, her boat passed quite close to the bachelors' camp and she seized the opportunity to have a good look at them. Half-relieved and half-disappointed, she wrote, 'I do not know what we expected to see, but we imagined it would be very gay, and would ring with shouts of merriment; instead of which we saw dismal-looking bachelors sitting in gloomy solitude in their boats looking bored to death.'

The missionaries in Srinagar, however, from whom not much was hidden, sometimes saw the bachelors in a very different light—one that, perhaps, would have matched the most vivid of Miss Parbury's secret imaginings. 'People seem to come here purposed to covenant themselves to all sensuality, and to leave what force of morality they have behind them in India', wrote Bishop French of Lahore, who came on a pastoral visit to Kashmir in 1871. Cecil Tyndale-Biscoe, the plucky missionary teacher, told how he was once called to the deathbed of 'a young Britisher who had succumbed to temptation, put in his way deliberately by the owner of the houseboat in which he lived, with the result that he contracted a dreadful disease and when I was called to minister to him, I found him paralyzed'. The Englishman died soon afterwards, but not before he had begged Tyndale-Biscoe to pass on his story as a warning to others. Tyndale-Biscoe needed no prompting to inveigh against what he called the 'floating houses of ill-fame', for his own brother Julian, on his arrival in Kashmir, had rented a houseboat from an agent only to find as he stepped aboard that two pretty Kashmiri girls were waiting there to greet him and his companion. 'Fortunately', wrote Tyndale-Biscoe, 'the agent was standing on the prow of the boat, so that when my brother's fist caught him squarely, the agent disappeared into the waters of the River Jhelum.'

The gentle young English missionary, Irene Petrie, who made good friends among the secular British community in Srinagar (including Geoffroy Millais, son of the well-known painter Sir John Millais, and an award-winning photographer himself), was driven by the behaviour of some of her compatriots to write, 'The worst thing of all in Kashmir is the conduct of some of the English people who find their way to this remote place. It is grievous to hear how the inquiring and intelligent natives point to them as the stumbling-blocks in the way of their accepting Christianity'.

For the missionaries themselves, life in Kashmir promised precious little *except* stumbling-blocks. The work was exhausting and interminable, and yet, in what was described as a 'decidedly encouraging' year, there were only eight Anglican converts, and from the point of view of health it was a cruelly taxing posting. Three of the missionaries to Kashmir died and many others were forced to abandon the struggle and return to England to recover their strength.

William Elmslie, a young Scottish doctor, was the first to be sent to the valley in 1865 by the Christian Missionary Society in London. He was also the first to die for it, but not before he had produced a Kashmiri–English dictionary which was invaluable to those who came afterwards. 'Poor perishing Kashmir, for whom I could weep all day', he wrote home to Scotland. His wife recorded that in one month, shortly before he died, Elmslie, working alone and under the most primitive conditions, had treated 1100 patients, and performed 70 operations.

Next to die was a woman doctor, Fanny Butler. She had been the first student to enrol at the London School of Medicine for Women, and had worked in India for six years before moving up to Kashmir in 1888 to start a clinic for women. On the day it opened only five terrified patients turned up. Within twelve months 5000 patients had been seen, but poor hard-working Fanny Butler was gone.

Five years later a young thing in her twenties, Irene Petrie, arrived full of enthusiasm and health, never considering for an instant that she would end in a grave even further from home than Dr Butler's in Srinagar. Irene Petrie taught in the mission schools and visited Kashmiri women in their homes to read to them from the Bible, sing hymns and teach them to read and knit. Shut away in their airless room, bored and very often sickly, these women looked forward to Miss Petrie's visits and crowded around her. In winter this became almost unbearable, for the Kashmiri habit is to keep warm by carrying a *kangri*—a small, insulated pot containing charcoal embers—close to the body, and each women held one under her grubby robes. Once or twice Miss Petrie was overcome by the stuffiness and smells and fainted clean away, but the only complaint she was ever heard to murmur was, 'Oh my dear Kashmiri women, why don't you wash?'

Miss Petrie lodged with the Tyndale-Biscoe family at Holton Cottage and her social life revolved around them and the other missionaries. In winter the isolation and monotony of their lives drove the secular community and the missionaries closer together than either side would have thought possible—or desirable—in summer. The most worldly people were only too grateful to while away an hour or so at the amateur concerts arranged by the missionaries in the Library, and even religious lectures like Dr Neve's on 'Recent Progress in New Testament Criticism' could produce a large turnout.

In 1896 there were great celebrations in missionary circles when Bishop Matthew of Lahore came to consecrate the three Anglican churches that had been built in Kashmir—two in Srinagar and one little one in Gulmarg for summer visitors. At

All Saints', where the Anglo-Indians (as the British in India were known at that time) worshipped, and where the British Resident read the lesson every Sunday, the service was sung in English, and Miss Petrie wrote that 'many said it was like a home church service'. She herself preferred the simple Urdu service in St Luke's, the tiny church which the missionaries had built for themselves and their few converts in the grounds of the hospital—Miss Petrie, a talented amateur painter, had raised the money to buy an organ for the church by selling her pictures. St Luke's is still there, unused now, and dilapidated. Its little brick spire will help visitors locate it in the crowded enclosure of what is now the T.B. hospital.

The last social events Miss Petrie attended were secular ones in celebration of Queen Victoria's Diamond Jubilee in June 1897. There was a fête in the Residency gardens, a grand durbar given by the maharaja—to which the missionaries were quite surprised to be invited—a review of the maharaja's troops, and a regatta put on by the boys of Tyndale-Biscoe School. But the most moving moment, wrote Miss Petrie, came during the military sports display when a telegram was handed to the Resident: 'It was the dear Queen's own message, and we read it with such a thrill, within half an hour of its despatch by her own hand.'

A week or so later, Miss Petrie set off on a trek to Ladakh where she contracted typhoid fever and died. By chance, the most senior missionary doctor, Arthur Neve, was in Leh when Miss Petrie was taken ill. He rushed to her bedside but could not save her—'Our stay in Ladakh was saddened by the illness and death of Miss Irene Petrie, a most charming and accomplished young lady . . .', he wrote. Irene Petrie is buried in the Moravian graveyard at Leh, and there is a plaque to her memory in what was her local church in London, St Mary Abbott's in Kensington.

Arthur Neve was a medical student in Edinburgh when he read the story of Dr Livingstone, and was inspired to become a missionary. His dream was to work in Africa and he was dismayed and disappointed when the Church Missionary Society posted him to Kashmir in 1882. He had a rather hair-raising ride into the valley—his suitcases fell off the pack horse and crashed down the precipice, bursting open as they fell, and his own horse nearly slipped off the narrow mountain path into the ravine. But when he reached the top of the hillside above Baramulla and saw the Kashmir valley laid out at his feet, he forgot all these difficulties. It was love at first sight, and for the next forty years Arthur Neve, and his brother Ernest who soon joined him, worked in the Mission Hospital at Srinagar which they rebuilt and expanded hugely, and where, he wrote, about twenty major and fifty minor operations were performed on most days. There were not many spare moments, but Arthur Neve spent any he had trekking in the mountains and exploring Kashmir and its neighbours. Eventually he put all his knowledge into a guide book called 'The Tourist's Guide to Kashmir, Ladakh, Skardo, Etc.' It is a book that would still be useful to a traveller today, for though the political upheavals of the twentieth century have closed some of the routes into Kashmir and have brought the great trading caravans to a halt, within the valley itself many of the paths and tracks and short-cuts that Neve described in detail are the same as they were in his day.

Of all these men and women who dedicated their lives to Kashmir, only one is remembered today—Cecil Tyndale-Biscoe, fellow-missionary and friend of the Neve brothers and Miss Petrie. Cecil Tyndale-Biscoe, born only six years after the Indian Mutiny of 1857, was in Kashmir for the heyday of the British empire, and was still there to watch its collapse. He was in Srinagar when the British Resident formally closed the Residency there in 1947.

'The great little man of Kashmir', as Canon Cecil Tyndale-Biscoe was called, had been born into a loving, prosperous upper-middle-class family in England. Their name was really plain Tyndale, but an uncle had promised to leave them his grand house on condition that they tagged Biscoe on to it. Despite this background, poor little pint-sized Cecil led a miserable childhood—first with a governess who punished him continually with the excuse that it was her duty to get the 'devil' out of him, and then at Bradfield school, which he described as 'five years in hell'. Boys were bullied mercilessly and assaulted and tortured by other boys, and Biscoe developed a lifelong sympathy for the weak and oppressed, and a loathing of violence—'knocking boys about makes me see red', he wrote. When he became a schoolmaster in Kashmir, his punishments never involved beatings, but were tailor-made to the crime: a boy who talked when the order for silence had been given would have to stand up in front of the class and continue to talk non-stop; a boy caught smoking would be made to puff cheroots.

Having read theology at Cambridge, he was ordained and volunteered for the Church Missionary Society. Like Arthur Neve, he longed to be sent to Africa and was bitterly disappointed to be posted to Kashmir instead. He knew nothing about the country, not even where it was, and had to pay a visit to the Royal Geographical Society to find out something about it.

On his arrival in Srinagar Biscoe lodged with the Neve brothers in Munshi Bagh, and next day went with the headmaster, whom he was to replace, to meet the boys in the mission school, a rickety wooden building overlooking the Jhelum river. The two hundred students were sitting on the floor wrapped in their long gowns, looking like so many bundles, and each one clutched a *kangri*, for it was winter. 'As I entered, the stench almost knocked me backwards', says Biscoe baldly. As he scanned the faces of his new pupils, he wrote afterwards, their expressions of slyness and evil took him right back to his early days at Bradfield. It was not long before Biscoe discovered that many of his pupils were boy prostitutes, and that any attractive ones who were not lived in daily peril from marauding bands of pimps. Once, such a gang invaded the school football pitch and tried to carry off some boys in the middle of a game. Biscoe taught his boys self-defence, particularly boxing, which they did in special cotton gloves because the Hindus disliked touching leather, and he set up houses of refuge all over the city where his boys could seek shelter if they were attacked.

The pupils in the mission school were all Hindus—no Muslims then sent their children to school as government service was closed to them anyway. Biscoe thought it ironic that he, who had wanted to go to Africa to combat slavery, had found himself instead in Kashmir, teaching slave-owners' sons—for, he remarked, the Hindus treated the Muslims little better than slaves. Biscoe found these youths so arrogant, selfish, dishonest and idle that he could not bring himself to refer to them as men but called them 'bipeds', and for more than half a century, he threw himself into the task of transforming his bipeds into MEN.

The first step, as he saw it, was less Bible study in stuffy classrooms and more *action*. This meant not only sports, but 'charity and self-sacrifice' in the community—firefighting, helping the poor and sick, rescuing animals, saving people from drowning. At first everything that Biscoe suggested was stubbornly resisted. The boys flatly refused to leave their classrooms, claiming that they had come to study and that they were far too nobly born to involve themselves in any of these undignified activities. Biscoe drove them out of the classrooms with a stick and bullied them into submission. The first time the boys played football, the pitch was guarded by teachers armed with sticks to prevent them running away, and boys who complained that they would be disgraced if people saw them learning to row were told to wear blankets over their heads while they practised. Pupils were 'persuaded' to learn to swim by the fact that non-swimmers' school fees were increased each year.

Eventually, the Biscoe Boys, as they were known, became famous for their sportsmanship and good works. They climbed mountains, swam across lakes, and staged popular weekly regattas in the school's fleet of boats. They fought fires with great bravery—where before they had watched and sniggered. They saved about forty people from drowning each year, they helped to clean the streets in cholera epidemics, and when the city flooded, as it often did, they went out in boats collecting the refugees marooned on roof-tops and up trees—a service for which professional boatmen demanded huge, unaffordable sums. In the bad floods of 1893, Biscoe and his new wife Blanche lost all their possessions—except their pet monkey and bear, which sat on the roof of the cookhouse. Biscoe also managed to rescue the church organ by heaving it with some help onto the roof of the church.

Eventually, the state school began to imitate the mission school's activities. Football, cricket, boating and other sports were introduced, and inter-school matches could then be held. A British sense of fair-play was not always evident in the opposing teams, however: once, in a tug-of-war between the mission and state schools, Biscoe suddenly noticed that the state school's rope was actually anchored around a raised flower bed; and on another occasion, he realized that there were thirteen players in their cricket team instead of eleven.

Canon Tyndale-Biscoe's extraordinary efforts often went unappreciated. There were those conservative British who believed that educating the 'native' was a bad thing, on the grounds that 'educated thieves were more dangerous than uneducated ones'. Some of his fellow missionaries were highly critical of his 'goings-on', and complained to London that instead of organizing fire-fighting parties in the city he should be praying and preaching and trying to convince a few more converts. Soft liberals such as E. M. Forster, who loathed the Kiplingesque ideal of manliness, were also extremely critical of Biscoe. Most dangerous of all, the Hindu establishment in Srinagar feared him for undermining their authority, exposing corruption, and working against vested interests. Three times there were threats against his life, and then Biscoe would sleep with a loaded revolver under his pillow and set up, in Holton Cottage, a Heath-Robinson type of alarm system made of strings and tins. But no threats deterred the little man in his determination to make men out of the Kashmiris, and there was no injustice or immorality that he would not tackle. When he saw that the weight of a grossly fat man was almost breaking the back of the pitifully thin donkey he was riding, Biscoe made him get off and carry the donkey instead. He championed the cause of Hindu widows who were often cruelly misused by their husbands' families. He set up a fund to help those in debt to extortionate moneylenders. He organized police raids on bookshops selling pornography—before becoming involved in this he first had to ask permission from the British Resident, who said warily: 'I have no objection, if you promise me that there shall be no row... Now, Mr Biscoe, mind you, there must be no row'. And a boy caught reading a dirty book at school was made to eat it—after Biscoe had checked with Dr Neve to see how much paper a person could consume without harm being done. Once, Biscoe was teaching a class when a member of his staff rushed in to tell him that a girl of thirteen had been snatched from her mother's side and taken to a house of ill-repute. Biscoe immediately sent his boys running off to surround the brothel and prevent anyone leaving. While he himself leapt onto his bicycle and pedalled furiously to fetch the police—'Prompt action is necessary in Kashmir lest one be checkmated', he wrote. He was not checkmated on that occasion, the girl was released and her kidnappers went to prison.

Biscoe had a good sense of style and drama—and public relations. The annual return of the maharaja from his winter quarters in Jammu was always a good excuse for the Biscoe boys to stage a superbly-orchestrated salute from their boats on the river in front of the school, and for a Viceroy's visit Biscoe organized his pupils into a complicated 'living welcome'—they climbed into a special frame suspended across the river and spelt the word WELCOME with their bodies.

In 1947, Biscoe's long reign came to an end—not because India was now independent, but because a new principal was taking over the school. Although Biscoe would have liked to stay on in Kashmir, 'it was thought that my presence might cause difficulties...'. Cecil and Blanche Tyndale-Biscoe were given the kind of send-off that their fifty-seven years of service deserved. Their car was pulled along by thirty of the school staff from Holton Cottage to the bus stand, with the school band marching ahead, and the road was lined on either side by cheering boys and old students. But after that, their journey was full of heartbreak, for when their bus crossed from Kashmir into what had suddenly become Pakistan, it was confronted by armed men searching for Hindu and Sikh travellers whom they were dragging out of their vehicles and shooting. They had killed

300 by the time the old couple arrived. And in Bombay, they heard 'the terrible news of the invasion of Kashmir by Afridi tribesmen and other devils who came up the 100 miles of road, bound for Srinagar with 500 lorries, burning, looting and slaughtering.' The next information they received—that the Indian Army had been flown up to repel the invaders—was more comforting, and then, just before their ship sailed, they heard that the maharaja had stepped down and that Sheikh Abdullah, the popular Muslim leader who had been imprisoned for agitating against the maharaja's government, was now in charge. This was positively good news, for Sheikh Abdullah had proved himself a staunch friend of the mission schools and had sent his own children there to be educated. Old Canon Tyndale-Biscoe died eighteen months later in Rhodesia, comforted by the knowledge that no less than fourteen of the ministers in Sheikh Abdullah's new government were old Biscoe Boys.

Epic scenes have been enacted on the small stage of Kashmir, mostly, it has to be said, ones of tragedy, conquest and invasion. But in 1850 a curious little domestic drama took place there, almost an English drawing-room comedy. The leading players happened to be the first two Western women ever to visit Kashmir, Honoria Lawrence and a Mrs Hervey, who had never met each other before, but were to do so in Srinagar in rather strange circumstances.

Honoria Lawrence we have already come across in Chapter I. She was the wife of Sir Henry Lawrence, who was then the uncrowned king of the Punjab (he was President of its British Board of Administration). Sir Henry had been one of the signatories to the treaty by which Britain handed Kashmir over to Maharaja Gulab Singh, and he knew the maharaja well. In 1850 he travelled up to Kashmir to keep an eye on things and to 'advise, help and control Gulab Singh, without affecting his authority—a delicate task'. Lady Lawrence joined him there early in July that year, and the Lawrences spent two and a half months in the valley, camping in various Mughal gardens around the Dal lake, and living in the maharaja's guest-house in Sheikh Bagh. They flourished in Kashmir. The cooler climate improved the family's health and spirits, and Lady Lawrence delighted in the sight of her two small children, Harry and Honey, playing in the shade of the giant chenar trees and splashing in the fountains and tanks of the Mughal summer houses: 'we only wish we could remain here for the rest of our Indian career', wrote Sir Henry to a friend. But about a month after his wife's arrival, Sir Henry had temporarily to abandon this scene of domestic bliss and trek off to inspect the maharaja's adjoining territory of Ladakh. On the way there he stumbled upon Mrs Hervey. He was travelling to Leh, Mrs Hervey was travelling *from* Leh to Srinagar, and their paths crossed at Dras, not far from the Kashmir frontier.

Not very much is known about Mrs Hervey, except that she was an exceptionally bold and attractive adventuress who had flung herself into travelling as a consolation and escape from an unhappy marriage. Mrs Hervey's travels were unbelievably daring, especially for a sheltered lady of those times. They took her into the remotest parts of Asia—through Tartary, Tibet, China and Kashmir, and including, as the title of her published journal stated, 'portions of Territory never before visited by Europeans'—let alone European *women*.

The idea of making this daunting journey came to her, Mrs Hervey says, when she was 'under the pressure of severe domestic affliction, which was paralysing every energy of mind and body'. It seems to have succeeded in anaesthetizing her pain—'in the wilds of the snowy Himalayas I almost forgot the world I had left and the memory of many bitter sorrows and trials was softened if not banished.'

The exact cause of Mrs Hervey's unhappiness remains a mystery, for she never wrote directly about her problems, and every name in her journal was reduced to an initial. 'E' seems to have been her husband—'so crafty and false', and 'O' her beloved son, about whom she felt particularly anguished: 'the poor boy is very unfairly treated', she writes. Even in the wildest places Mrs Hervey was often busy writing and sending off affidavits concerning her husband's treatment of the boy. Presumably, Mrs Hervey's husband was an army officer, for she wrote feelingly of army wives as 'galley-slaves', and in one outburst in French declared that of all types of slavery, military life was the most onerous.

Most of her remarkable journey was made alone, but while passing through Kulu Mrs Hervey re-encountered a certain Captain H– who was Assistant Commissioner there and an old acquaintance of hers. Captain H– expressed himself worried about the dangers surrounding a woman travelling alone and begged to be allowed to accompany Mrs Hervey for part of her ongoing journey. She agreed, not because she feared danger but because she found him a pleasant, if trifle stubborn, companion: 'He will not listen to my sage advice', she writes after a disagreement between them, 'so it is of no use my offering him any'. Just as stubborn herself, Mrs Hervey ignored any suggestion that a young married woman travelling alone with a man who was not her husband would cause scandal.

So Captain H– and Mrs Hervey set out, travelling alongside each other rather than together, for he and she both had their own equipment, foodstuffs and servants. However, the hazards of the journey meant that quite often the porters carrying one or other of their camps would fall behind, or get lost, or turn up at the wrong rendezvous, and when that happened Mrs Hervey and Captain H– would have to pool what resources they were left with. They were sharing a camp, in fact, at the time they met Sir Henry Lawrence on the road to Dras, and this did not endear Captain H– to Sir Henry, who thought it quite shocking that Mrs Hervey should be compromised in this way, and that Captain H– ought to behave like a gentleman and return to his duties in Kulu immediately. Sir Henry sent a report on his meeting with the errant couple to Captain H–'s superior which spelt out the facts: 'The night before we met . . . she had a bed and he had none; and they had only one tent up. The same thing would have happened last night, had we not given them shelter.'

Sir Henry was far less critical of Mrs Hervey herself. He thought her 'wayward and headstrong, with very lax notions on many subjects', but he could not help admiring her courage and stamina: 'She had had no dinner the day before, no bedclothes at night . . . she has been over some of the worst passes: and proposes now to go on from Kashmir, across the hills to Kunawar and Simla. I advised her against it, and would have prevented her, could I have done so without using violence.'

Sir Henry's assistant, Mr Hodson, was obviously rather captivated by the spirited Mrs Hervey, and lent her his own tent for the night. He described her as

a very pretty creature, gifted with indomitable energy and endurance—except for her husband, whom she can't endure, and therefore travels alone. For three months she has been pony-riding through country few men would care to traverse, over formidable passes and across the wildest deserts in Asia. For twenty days she was in Tibet, without seeing a human habitation; often without food or bedding. When we met her, she was over sixteen miles from her tents, rain and darkness coming on apace. So we persuaded her to stop at our encampment. I gave her my tent and cot and acted lady's maid; supplied her with warm stockings, towels, brushes, etc. Then we sat down to dinner; and a pleasanter evening I never spent.

Mrs Hervey enjoyed herself too. She writes in her journal in her typically wry way—'I was agreeably disappointed in Sir Henry Lawrence. He appears a kind and amiable man.' Next morning they all breakfasted together and then parted company—Sir Henry and Mr Hodson to continue their journey to Leh, and Captain H– and Mrs Hervey theirs to Kashmir.

Mrs Hervey had told Sir Henry that she would like to see the Wular lake in Kashmir, and so he had devised a roundabout route for the couple which took them into Kashmir via Sonamarg, across the valley to Manasbal lake, and then up to the Wular lake and down again to Srinagar. Sir Henry seemed to have forgotten that it was August and the height of the mosquito season—the worst possible time of year to be exploring the lakes—and the pair spent a wretched, sleepless first night in Kashmir beside the Manasbal lake. Nonetheless, next day Mrs Hervey was determined to proceed with the plan, and hired a boat to take her to the Wular lake. Captain H– apparently could not find a boat to hire, and went straight on to Srinagar. Mrs Hervey spent two awful days and nights on the Wular lake. Dramatic storms, one of the few things she was afraid of, delayed them for hours, her bed was always damp, if not soaking wet, the water was rough and the mosquitoes relentless. 'They will shortly deprive me of my senses, or drive me into a fever', she says.' 'They bite me morning, noon and night, and drive me half mad'. It was with relief that early on the third morning Mrs Hervey saw Colonel Henry Steinbach, an English officer in the maharaja's army, coming across the lake to meet her and escort her to a camping ground in the Nishat garden. There she installed herself under the trees of the ladies' terrace at the back of the garden and, delighted to be off the water and away from the mosquitoes, wrote: 'This place—the Nishat Bagh—is a lovely spot; trees and water, and ruins in wild beauty, meet the eye on every side'. Captain H– was also camping at Nishat, and that evening the couple went on an excursion to the Shalimar garden, famous to both of them because of Thomas Moore's poem 'Lalla Rookh'. The night air brought out a few mosquitoes, but Shalimar lived up to its romantic reputation, and Mrs Hervey wrote in her journal that night:

> The palace is in ruins, but the gardens, indeed the whole of the grounds of Shalimar, are very lovely. The fountains were playing in honour of our visit, and the last declining rays of the sun gilded the black marble pillars of the palace . . . the very wildness and desolate beauty of the spot had charms for me.

Next day, Mrs Hervey discovered that Nishat was frustratingly far from the city, and she became impatient to move. There was no house free in Sheikh Bagh, where the Lawrence's were, but Colonel Steinbach arranged for her to take over a house on the river in Munshi Bagh, and there Mrs Hervey went, to relax and rest as well as explore, for she was still tired from the long trek from Leh and the sleepless nights she had spent on the lakes—'Even my face looks worn and wasted', she says.

Mrs Hervey does not mention it in her journal, but Lady

Lawrence does in hers—that very soon after this Captain H– went alone to call on Lady Lawrence. She disliked him even more than her husband had done, and she was deeply embarrassed by what she described as the 'very unnecessary particulars' he gave her concerning Mrs Hervey's virtue. Lady Lawrence was not one to mince words, and she told Captain H– frankly that if he was really only Mrs Hervey's *friend*, as he claimed to be, then he would stop compromising her and return to Kulu. Her talk seems to have had quite the wrong effect, for in the middle of that very night she received an urgent note from Mrs Hervey begging for protection as she had been 'grossly insulted' by Captain H–.

Why Captain H– should have chosen this moment in Srinagar to press his attentions on Mrs Hervey when they had been alone together—apparently quite happily—for weeks beforehand, is a mystery, and it seems likely that there was more to the story. Perhaps Mrs Hervey had not fully realized when she set out with Captain H– what it would be like to be disapproved of by people she respected like the Lawrences. The thought of Sir Henry catching them sharing a tent and Honoria Lawrence not receiving her in Srinagar was probably deeply humiliating to her. Indeed, it may have been Mrs Hervey who sent Captain H– to call on Lady Lawrence to try and redeem the situation by making all those protestations about her virtue. When Captain H– reported that his mission had failed, Mrs Hervey may have decided that the only way left of proving that she was not already having an affair with Captain H– was by pretending that he was pressing her to start one—perhaps they arranged it between them, he agreeing to sacrifice his honour in order to redeem hers.

We shall never know why Mrs Hervey wrote her plea for help, but we do know that Lady Lawrence acted generously: 'Dear Madam', she wrote immediately,

On receipt of such a note as yours, there is but one step for me to take—to beg you will at once come here and place yourself under my care, where you will be as safe as if you were my daughter. I have ordered the boat that is at my disposal, and one of my servants will attend. Pray do not lose a moment in coming. It is exactly night when your note has arrived. Do not delay to pack. I can send for your baggage in the morning. Yours sincerely, H. Lawrence.

The strange thing is that Mrs Hervey did not appear in answer to this invitation, but sent back a polite excuse. Maybe she panicked at what she had done and at what she would now have to do—leave Munshi Bagh and Captain H– and go to live respectably with Lady Lawrence—or perhaps she was just postponing the parting from her lover.

Unabashed by the lack of response, Lady Lawrence wrote again next morning:

I am sorry you did not come over last night. A bed was prepared, and you would not have put me to any inconvenience. Let me urge you to come today. At the same time, as one old enough to be your mother, let me speak frankly but kindly. Had you come to Kashmir alone, I should at once have asked you to be our guest, but I could not countenance what I felt to be very wrong in your conduct.... A young married woman so unhappy as to be separated from her husband... should not wilfully expose herself to scandal, insult and temptation. Yesterday you would probably not have listened to these words; today I hope you will consider them. If so, I shall gladly ask you to be my guest—and give you any help I can in returning to your brother, or whoever else may be your proper protector... Do not cast away my letter as harsh. Indeed it is not in that temper that I write. Pray come today, whether Captain H– leaves you or not. Alone, you cannot feel safe from a repetition of insult. I have a room ready, and I dine at two o'clock, by which time I hope you will be here.

Once again Mrs Hervey failed to respond but sent more excuses—including the information that she would probably be leaving Kashmir next day anyway. Lady Lawrence wrote a third time: 'I repeat you are heartily welcome, and I hope you will not hurry away. I will send my boat at twelve to bring you... Let nothing prevent you from coming to me. Trust me to judge for myself of you... It would be very foolish of you to leave Kashmir tomorrow. I will at once send my boat for you.'

These letters from Lady Lawrence appeared in a biography

of Honoria Lawrence written by Maud Diver and published in 1936. It seems that Mrs Diver had not read Mrs Hervey's journal, for her version of the story was that 'the wayward one' (as she referred to Mrs Hervey) never took up Lady Lawrence's invitation but left Kashmir soon after this exchange of notes. Mrs Hervey's journal, however, makes it plain that she *did* go and stay with Lady Lawrence. She gives no explanations, mentioning neither being insulted by Captain H– nor her plea for protection. But two nights after recording her move to Munshi Bagh she says: 'I left Moonshee-ka-Bâgh this forenoon, and accompanied Lady Lawrence to her mansion here: a fine, large house, made over to her *pro tempore* by his 'Highness'. She came to fetch me in the Rajah's large barge, and I felt indebted for her kindness to a stranger.'

Captain H– now drops out of the story, and his name is not mentioned again, except briefly a day or two later when Mrs Hervey writes without emotion: 'Captain H– left Kashmir for Kangra by horse-dak this morning'. The 'we' of so many adventures and excursions recorded in Mrs Hervey's journal now becomes a solitary 'I', and although she continued to explore diligently her style becomes less lively, a little bleak, as though a light had gone out of her life.

Her gloomy mood was not improved on the day when, just outside the city, she passed two wooden cages hanging from gibbets, each one containing human remains. 'Their clothes were still on them, and these ghastly skeletons looked very revolting, bleaching in the bright sunlight. They have been hanging for two or three *years*; their crime was murder'.

A few days later, Mrs Hervey completed the preparations for her onward journey through the mountains to Simla. There had been devastating floods in the areas she planned to travel through, and her friends in Srinagar begged her to reconsider her plans. But Mrs Hervey's determined mind was made up, and in the end even Lady Lawrence was persuaded to help. She lent Mrs Hervey the state barge—which the maharaja had put at the disposal of the Lawrences—so that she could make the river journey from Srinagar to Anantnag (which used to be called

Islamabad) in comfort. At Anantnag Mrs Hervey paused only long enough to admire the 'graceful trellis and lattice work' on the houses, and then she was off again on horseback to make her way through the mountain passes of south-east Kashmir into Kishtwar and beyond.

Less than a year later, Mrs Hervey was back in the valley. It seems rather typical of her that she reappeared in the exact opposite corner of Kashmir to the one by which she had left it nine months before, marching in for her second visit from Punch, west of Kashmir. At Baramula she hired a boat in which to make the journey to Srinagar, for she had decided, once again, to approach the capital by way of the Wular and Manasbal lakes. The previous year mosquitoes had made life on these very lakes unendurable, but this time around it was May, not August, and too early for the torturers. Mrs Hervey was able to concentrate on her surroundings: 'I never saw such lovely scenery as delights the eye while sailing lazily along this part of the beautiful Vale of Kashmir', she writes happily.

She re-visited the place where she and Captain H– had spent the night, and saw traces of their camp still imprinted in the grass. Nostalgically, she plucked a sprig of leaves from one of the great chenar trees that shaded the site and pressed it in her herbal as a souvenir. From there she gazed across the still water of the lake to the fine trees and tumbledown terraces of Jarogha Bagh, once the garden palace of Nur Jahan, and thought what a handsome old ruin it made.

That evening near Sopor she dined under a tree and then, in a contented mood, wrote up the day in her journal: 'There were no maddening insects to destroy the pleasing romance and beauty of the scene and hour; and I could uninterruptedly feel, "how beautiful is Kashmir".' However, it did not end on quite such a tranquil note, as Mrs Hervey revealed in a footnote written later. She retired to her boat for the night, completed her *toilette*, and then, space being very restricted, she perched on the side of the boat in order to pull back the curtains around her bed so that she could climb in. In doing this she somehow overbalanced and toppled backwards into the icy water of the lake and had to be fished out by her astonished servants. It was, she commented quite calmly considering her ordeal: 'a very cold bath at a very inopportune time'.

Unscathed, Mrs Hervey arrived in Srinagar next day and tried to book herself a house in Munshi Bagh. To her disgust she found that every one of them was already taken, for, as she soon discovered, the town was full of British visitors. Less than one year before, Mrs Hervey had missed the distinction of being the very *first* white woman in Kashmir by only a matter of days. Now she felt all the contempt of the old hand for newcomers, and was positively bursting with righteous indignation at the ignorance and bad manners of these shallow 'tourists'. '*Mon Dieu, quel accés de Vandals et de Goths*!' she says acidly. 'Every *nullah* and gully is crowded with Philistinian faces, and each beardless or bearded cockney considers himself entitled to desecrate this sweet valley of romance'. She did not mean that Kashmir had been overrun by Londoners born within hearing of the Bow bells—'I use the word 'Cockney', she explains, 'to represent a person who knows but little of travelling, *de facto*, or of its aims'.

Mrs Hervey quickly removed herself to an old haunt, Nishat Bagh, where, this time, she camped in the summer house by the lake, rather than on the ladies' terrace at the back of the garden. She liked this pavilion with its exquisitely carved pillars and doors and windows, and its lovely view over the lake—there was no road separating the garden from the water in those days. The noise of the fountains and cascades soothed her ruffled spirits and she writes thankfully: 'I am very glad to be out of the way of those English hosts, and, in this sylvan retreat, fancy myself still in a secluded place, far removed from the impertinent eyes of my country folk.' Then she added a few favourite lines of poetry in praise of being alone, rather touchingly underlining the last words to make her feelings absolutely clear:

> There is a pleasure in the pathless woods,
> There is a rapture on the lonely shore,
> *There is society where none intrudes.*

Although Mrs Hervey so greatly despised the visitors to

Kashmir who came after her, she felt nothing but admiration for the adventurers who had preceded her to Kashmir in times more difficult and dangerous—with the exception of Baron Hügel, whose pompous style, not surprisingly, irritated her. On this visit to Kashmir Mrs Hervey made two pilgrimages to places that those earlier travellers had known and liked, and which were familiar to her through their descriptions. First, she went to the Char Chenar island where Victor Jacquemont had once camped, and where Baron Hügel, Godfrey Vigne and Dr Henderson had arranged to put the marble slab commemorating their visit. She was very disappointed in the island: 'A few ruined pillars, and the foundation of a house, was all that was discernible in a heap of rubbish and jungle.' Next, she went to inspect Dilawar Khan's garden in the city of Srinagar. This had been the camp site allotted to Jacquemont, Hügel, Vigne and Henderson—all of whom had enjoyed it—but Mrs Hervey was again disappointed: 'I do not like the situation of Dilawur Khan Bagh. There is a dreary and desolate aspect in the swamp, which faces the principal latticed window, or one entire side of the building.'

As Mrs Hervey was paddled through the waterways of Srinagar and its suburbs, she was struck by how little had changed since her last visit. Even the two skeletons in the wooden cages were still there, horribly the worse for wear. 'The only strange and disagreeably new feature was the frequent passing of *white* faces', she says crossly.

That evening, her last one in Srinagar, Mrs Hervey's servants accidentally broke her precious thermometer and she was almost as upset as she had been about the death of her dog the year before: 'I have been trying to *cobble* my poor broken thermometer, but all in vain; half the quicksilver has already vanished. In this benighted land a thermometer is unknown, and no money can purchase one. I am very much provoked at my loss, as I wished to register the temperature regularly in the wilds of Tibet. Well, I must do without, *c'est tout*.'

Next day, after less than a week in Srinagar, the restless Mrs Hervey set off once again for Anantnag, from where she planned to ride on to Verinag and see Jahangir's garden, and then on out of Kashmir by the Banihal pass into Jammu. All went well as far as Anantnag, where this year she noticed for the first time the attractive gabba rugs which are a speciality of the town. 'Pretty puttee patchwork', she calls them (*puttee* or *puttoo* was—is—the name of the local homespun wool). But from there on the journey became a nightmare. It rained continuously and their route lay across flooded rice fields into which Mrs Hervey's pony kept stumbling from the narrow raised paths they followed, so that she was not only soaking wet but covered with mud as well. Progress took so long that darkness fell and the dried grass torches of the Kashmiri coolies were soon burnt up; thunder rolled around the valley threateningly and flashes of lightning lit up the sodden scene and, Mrs Hervey wrote later, she 'could have cried like a baby'. By the time they reached Verinag she was sick and feverish—but for this she was grateful for it meant she did not feel the cold.

In the morning, however, Mrs Hervey awoke feeling well again, full of energy, and ready to face the climb out of the valley. From the top of the Banihal pass she looked back across Kashmir for the last time. The sun was shining and all the previous day's miseries were forgotten: 'How calm, how lovely seemed the smiling vale', she says, already feeling romantic about the place. For, like all such journeys, every stage of Mrs Hervey's was far pleasanter in retrospect than it was to live through, and every country seemed that much more attractive once it had been left behind.

Mrs Hervey published her journals in 1853, and, although they filled three fat volumes, she herself remains a mysterious figure. We know that she returned to England, for the sentence quoted at the beginning of this chapter ends her third volume. But what became of her—or 'E' or 'O' or Captain H– does not seem to be recorded. This lack of information is all the more surprising in view of the fact that she also wrote two novels, both highly-coloured romances full of orphans, cruel husbands and hopeless love that conquers all in the end.

In courage and endurance Mrs Hervey was a match for any

of the better-known Victorian travellers, and yet neither her journey nor her journal seems to have received much acclaim. One cannot help wondering why. Was it because of her 'wayward' morals, or the fact that she did not have the snob-appeal of some of the other women travellers in the East? Emily Eden, for instance, the sister of Lord Auckland (Governor-General of India), whose two books of travels—one published before, and one after Mrs Hervey's—were far more popular.

As Mrs Hervey had observed with such irritation, it was not, by 1851, particularly unusual to find Europeans in Kashmir. The valley was reasonably stable, its new ruler, the maharaja, welcoming, and though the journey there was still tiresome and uncomfortable, the element of danger had been removed since Britain had defeated the Sikhs and annexed the Punjab. The secret garden that only a handful of Europeans had glimpsed—sometimes risking their very lives to do so—was now open to the public.

Indeed, by the end of the nineteenth century it was fashionable to complain that Kashmir was 'spoilt', and Sir Francis Younghusband, who had known the valley in more rugged times, lamented how tame it had all become. To enter the Paradise of the Indies was no longer a daring enough feat to be an end in itself.

But Kashmir slipped effortlessly into a new role. Instead of a challenge, the valley became a resort, offering not only an escape from the terrible heat and frightening epidemics of disease that stalked the plains of India in summer, but, as well, views for the painter, mountains for the climber, ruins for the archaeologist, bracing air for the convalescent, plants for the botanist, and—as soon as two or more were gathered together—a lively 'season' for the socialite. And what nature had not provided—tennis courts, cricket pitches, golf links and so on, the British soon did, for as Sir Francis wrote rather charmingly, 'Man cannot live forever on walks . . . His soul yearns for a ball of some kind whether it be a polo ball, a cricket ball, a tennis ball, a golf ball, or even a croquet ball. Unless he has a ball of some description to play with he is never really happy.'

The British, just like their predecessors in Kashmir, the Mughals, seem to have been happiest of all, however, when hunting and killing things, and shooting and fishing were by far the greatest of Kashmir's attractions, at any rate in the early days. By the 1900s, the sahibs had thoroughly organized the sport for the maharajas. Licences were insisted upon, game preserves set aside, and a limit was set on how many heads of ibex, markhor, stag, black bear, brown bear and leopard could be killed. A trout hatchery was set up at Harwan, just beyond the Shalimar garden, and stocked with fine English trout supplied by the Duke of Bedford. These were released into Kashmir's rivers and streams to make it even more of a fisherman's paradise.

A perk that went with the job of British Resident in Kashmir was the right to the shooting on the Hokrar lake and marsh—this was a traditional gift from the maharaja. Francis Younghusband describes a great duck shoot he organised there, and how he and his guests crouched hidden in the reeds of the lake—themselves the targets for clouds of mosquitoes and gnats—while the sun beat down and the air shimmered with heat as they waited for the farthest gun to take aim and shatter the silence:

> At last a distant shot was heard, and then a suppressed roar, as of breakers on a far-off shore. Then from the direction of the shot a black cloud arose and advanced rapidly upon us. The roar increased, and in a few seconds the whole sky was covered with a swirling, swishing, whizzing flight of ducks. Thousands and thousands of them: flashing past from right to left, from left to right, backwards and forwards, forwards and backwards, in bewildering multitudes. For the moment, one's breath was absolutely taken away. There was such a swish and swirr, it was impossible to aim. Then, as the first wild rush was over, it became easier to be deliberate, and duck after duck fell to my companions' guns.

It was not unusual at a shoot such as this one for the party to bag more than 1500 ducks. And yet Younghusband still mourned the days of his first visit to Kashmir in 1887 when there had been absolutely no rules and regulations governing the sport at all, and a man could shoot as much as he liked wherever he chose.

The necessity for curbing visitors' trigger-fingers was vividly, though unintentionally, demonstrated by Captain Knight, who described, in a book of holiday reminiscences published in 1863, how he shot a big brown bear in Kashmir, but then lost interest and could not be bothered to walk down the hill to inspect the great animal whose life he had snatched away so pointlessly. One cannot dredge up much sympathy for the Captain when, a little later on, he rounds a corner in the mountains and is horrified to find himself face to face with the body of a local criminal hanging from a tree.

In a roundabout way, shooting literally changed the face of Kashmir. In the 1880s an Englishman called M. T. Kennard was so passionately addicted to the sport that he used to spend the winter months in Kashmir hunting everything from duck to ibex. The maharaja had decreed that no foreigner could buy land in Kashmir. Houses to rent were in short supply or were unsuitable in winter and it was too cold to camp out for any great length of time, so Mr Kennard, who was noted for his scientific turn of mind, designed himself a houseboat. This gave him a comfortable place of his own to return to every year without breaking any rules. The idea caught on, and twenty years later, when Sir Francis Younghusband was Resident, he wrote that the houseboats could be 'numbered by hundreds'.

A modern Kashmiri houseboat is a huge thing with, perhaps, three large double bedrooms and bathrooms, an enormous sitting-room, a dining-room and a verandah—all furnished with large pieces of locally-made carved wooden furniture, carpets and chandeliers. In Sir Francis' day they were smaller, but there was still room for 'a couple of sitting-rooms with fireplaces'—for although most of the boats were only let in summer ('from Rs 70 to Rs 100 per mensem for the season') some were year-round homes for Europeans—like Paul Scott's fictional character, Lady Manners, in the *Jewel in the Crown*.

Nostalgia for England inspired the designs of some of the early boats, which looked just like floating English country cottages with shingled roofs and steep gables and even black and white 'Tudor' timbering. And it was not long before little English gardens had been created around their moorings so that roses and hollyhocks, lupins and delphiniums bloomed along the Jhelum river.

Kennard's houseboat was, basically, only a more sophisticated and solid version of the traditional Kashmiri boat called a *doonga*—a long narrow craft with living accommodation made of straw matting tacked to a wooden frame. *Doongas* had been paddled around Kashmir's waterways since time immemorial, and they remained popular with those Europeans who could not afford the more expensive houseboats. In fact, they had two advantages over the houseboats—they were cooler in summer and were lighter, and therefore, much more manoeuvrable and mobile for travelling Kashmir's waterways. Their main disadvantage, apart from limited space, was that they were not burglar-proof. A thief could simply put his hands under the straw matting blinds and help himself.

An Englishman, A. Petrocokino, hired a *doonga* for a holiday in Kashmir in 1920 and described it as looking like 'a long hayrick on a punt, or an elongated Noah's Ark made of reeds'. A separate *doonga* travelling alongside acted as the kitchen and as housing for the boatman and his family and the other servants. 'This way of travelling is most luxurious', wrote Petrocokino, 'one sits in an easy chair in the porch of the dunga, gliding through the water with one's house, dependants and all one's belongings about one . . .'.

These, and other little domestic details, were revealed in *Three Weeks in a Houseboat*, the book Petrocokino wrote when his holiday in Kashmir was over. Publishers must have been a good deal more tolerant in those days, for no matter how prosaic or dull their experiences, anyone who set foot in the Kashmir valley seems to have been irresistibly drawn to jotting down their memories the moment they reached home, and sending them off for publication. However, even the least worthy of these books makes fascinating reading today—as much for the glimpses it gives of the British, as for what it tells us about Kashmir.

The most practical of the early books was a slim volume which looked at the valley from the down-to-earth point of view of an English housewife. *Cashmir en Famille* was written by a Mrs Burrows who, having spent four happy and healthy months with her husband and young baby in Kashmir in 1894, felt that irresistible urge to pass on 'some practical hints'.

'Life in Cashmir is very rough', warns Mrs Burrows: 'a picnic existence from start to finish—so only the simplest and strongest of everything is necessary, from the folding camp table to the sahib's shirt.' Mrs Burrows assures her readers that the Kashmiri washermen would reduce their clothes to shreds and patches, and that rats would probably eat their best linen tablecloths:

> As something of this kind is sure to happen, it is wise to be prepared and to buy some cheap table linens, etc., at the stores . . . it will be quite good enough for Cashmir. Everyone is in the same situation and quite understands and enjoys the odd make-shifts one has to go in for while travelling in these far away regions.

Mrs Burrows goes on to suggest that table silver was probably better left behind, 'for it is very likely that, while on the boat, a few spoons and forks will fall into the river while being carried from boat to boat; in fact, in one case we know of, the whole—servant, silver and all fell in . . .'. The Burrows themselves played safe by taking only 'common stag-horn knives' and enamelled plates, dishes, cups and tumblers. Mrs Burrows' wardrobe was as no-nonsense and hardy as her household effects, and included two strong tweed skirts turned up with leather at the hem, and two pairs of very thick waterproof boots with nails. It was a waste of time to take anything 'flouncy', for 'no smart gowns are required anywhere in Cashmir'. Both men and women were urged to order homespun wool suits in Srinagar as these cost only seven rupees, and were 'certainly good enough for use when marching in Cashmir and at other times if the wearer is not too particular as to the build of his garments.'

The Burrows arrived in Srinagar in April, when it was still

windy and wet, and they did not care for the place at all. For future travellers' sakes they hoped the rumours of a hotel being built were true, 'though the idea is unpleasing from an artistic point of view'. They moored their houseboat in town only long enough to order their *puttoo* suits and works of art and provisions, and then they set off to explore the valley. Starting at the Manasbal lake, where they camped on the terraces of Nur Jahan's garden for two blissful weeks, they moved slowly to Pahalgam, not forgetting, *en route*, to collect a large supply of the biscuits which were the speciality of Pampor and which Mrs Burrows recommends as being not only delicious but long-lasting as well.

At the end of June they made their way to the hill-station of Gulmarg, where they had hired one of the little wooden huts recently built for visitors. Theirs was one of the smallest with only two rooms and a bathroom, but Mr and Mrs Burrows made themselves comfortable with a tent pitched in the garden. The state carpenter came and put some hooks and shelves in the hut and made a plank floor for the tent, and Mrs Burrows laid down straw matting with pretty embroidered felt rugs on top, bought some wicker chairs and tables 'for a few annas' and ornamented their sitting-room with flower- and fern-filled *kangri* baskets (the real purpose of these was to insulate the charcoal warming pots which Kashmiris held under their clothes in winter). The memsahibs surreptitiously eyed one another's efforts at interior decoration, trying to pick up ideas for furnishing their own bare huts—'I have seen a wonderfully effective room trimmed with nothing but cheap muslim bought and dyed in Gulmarg for a few annas a yard', enthuses Mrs Burrows.

After a couple of weeks, Gulmarg began to pall and the Burrows were glad to get going again. 'It is difficult always to realize it', writes Mrs Burrows, 'but the charm of Cashmir lies in the marching about to fresh scenes and through fresh experiences, and when you find yourself settling down at Gulmarg into something like the ordinary routine life of an Indian hill station, you are apt to be bored . . .'.

The Burrows arrived back at their home in Bombay in September, almost exactly six months after they had set out so apprehensively for the unknown, and Mrs Burrows sums up the Kashmir experience for her readers: 'it is a trip that we can heartily recommend to all those sensible beings who think of their health, pleasure and pocket before frivolity'.

The most lavish of the holiday memoirs was written by a lady called Florence Parbury and published in 1909—though her visit to Kashmir took place in 1902. It was called *Emerald Set with Pearls* and the weighty cover was decorated with a colour picture of a huge cabuchon emerald surrounded by gleaming pearls. The contents are a peculiar hotch-potch of Miss Parbury's reminiscences interspersed with her own rather amateurish water colours and some photographs, plus Thomas Moore's poem 'Lalla Rookh', set to music, with illustrations of the costumes worn by the Russian and German royal families when they staged the poem in Berlin in 1822.

Florence Parbury explored Kashmir with her mother, who usually went for their 'walks' carried on a type of litter and they must have made an eccentric couple: 'Often I would find my mother enraptured with the beauty of some fresh scene, dreamily murmuring, "Beautiful, Oh! beautiful"; and I, too, would gasp out "Beautiful", much to the astonishment of the coolies who carried her dandy, and were usually too intent on finding a sure foothold to notice mere scenery', writes Miss Parbury. Once, in a meadow in the Lidder valley, she and her mother were quite carried away by their surroundings—'It was a paradise of flowers . . . we felt so much in touch with Nature, that, as there was no one in sight, we took out our guitars and sang some of our old songs.' It is rather delightful to think of these two very proper English ladies sitting in a field under a mountain in far-away Kashmir strumming on their guitars and singing out of sheer happiness and *joie de vivre*. Not surprisingly, they soon attracted an audience. By the time they finished their impromptu concert, there were seventy-five Kashmiri tribesmen crouched around them, fascinated and mystified. But this story gives a misleading picture of Miss Parbury, for she was by no means always so romantic and could be quite hard-headed at times, not to mention humorous. She certainly did not conceal the fact that there were definite drawbacks to a holiday in Kashmir—the quarrelling of servants, the stealing of garments and jewellery, the cruelty to animals and 'the cook's preference for the cooking of your food with river water'. The quality of the food obviously left a great deal to be desired, for Miss Parbury jokes that it would be more appropriate if the stringy chickens were boiled for two weeks rather than two hours, and has this to say on the subject of meat:

One of the drawbacks of the country from the English point of view is that a cow is a very holy and sacred creature: therefore, it cannot be killed to provide the 'Roast Beef of Old England' for the hungry traveller. It is a privation with which only visitors to Kashmir can sympathize and understand. I have heard of people ordering thick steaks of mutton for dinner, vainly hoping that their imagination, with the addition of a large supply of mustard, would make them believe they were eating beef steak; and of others who mixed a sauce of horse-radish powder and cream, and tried their luck at 'roast-beef-mutton'. We were so tired of mutton that we wired on to the hotel in Murree before our return journey to have beef of any kind ready for us upon our arrival.

The notorious Kashmiri vendors amused her, and she relates how one day, while crossing the lake in a *shikara*, she was hailed by one shouting 'Buy from me, Mem-Sahib, I am Cheap John'—at which point another vendor paddled up furiously calling out 'No Mem-Sahib, he is Cheat John'. Another vendor proudly showed her his book of references, supposedly from satisfied customers, in which she read: 'This is the biggest thief and scoundrel generally that I have ever met.' Miss Parbury seems to be the earliest traveller to record a meeting with an early version of that most famous of all dealers, Suffering Moses. He arrived on her boat and announced himself to her with the words 'All sahibs buy from me, my name is Suffering Moses.' Miss Parbury says: 'I felt inclined to call myself "Suffering Florence" by the time he had finally packed up his things and left me the dazed possessor of several things I did not want.'

Some years later, when she was assembling her book in England, Miss Parbury wrote of her longing for the keen bracing air and the magnificent scenery of Kashmir, but, she confessed, there was an impediment to her returning 'for another voyage through the Red Sea would kill me. . . .' Most British tourists at that time, however, seemed to find the new road from Rawalpindi to Kashmir more off-putting than the whole of the rest of the journey to India—certainly almost everyone had a bad word to say about it. The Jhelum Valley Cart Road, as it was called, was completed towards the end of the last century, and it meant that for the first time in history, travellers could be transported to and from the valley in wheeled vehicles rather than having to walk or ride—but this was still far from being a pleasant experience. The drivers of the carts and carriages were unskilled and impatient, and the careless cruelty they inflicted on their horses was so upsetting to the British that it was not long before two inspection posts were set up on the road by the Society for the Prevention of Cruelty to Animals. The road itself was terrifyingly precipitous in parts, with a surface that was bumpy at the best of times and impassable at the worst.

Marion Doughty made the journey to Kashmir in a *tonga* in 1901, and wrote that the passengers were so thrown about that 'I and my partner nearly succeeded in changing places like tennis balls'. Miss Doughty was not only uncomfortable on her journey, she was frightened; and, like a nervous air traveller today, had to keep reminding herself of all the people who *had* survived the journey. Nonetheless, on the precipitous mountain road, in the dark and rain, and at the mercy of a reckless driver and uncontrollable horses, there were moments of sheer terror, and Miss Doughty was astonished when they reached the journey's end unscathed:

I have been run away with, I have hunted on wheels in Ireland, and I have raced in a two-wheeled cart for a wager, but nothing had prepared me for that wild thirty-mile burst to Baramula. I have not ridden a mad bull through space, nor been dropped from the moon in a barrel with iron coigns, but I imagine a combination of the two would approximate our pace and progress.

These memories and reflections appear in *Afoot Through the Kashmir Valley*—Miss Doughty's contribution to the ever-increasing mountain of books on that country. In it she recommend Kashmir wholeheartedly, not only to fellow travellers but to any Englishman looking for a job out of the rut. Many, she observes, had already found employment as architects, road contractors, builders and engineers, but there were still openings for men with some 'technical training', and no one need be put off by the idea of working in such a remote spot: 'Of late various discomforts that tended to discourage visitors to Srinagar have been removed.' In fact, by then—1901—the European quarter of Srinagar boasted all the amenities, including a couple of agencies run by retired army officers which could supply everything a traveller needed. Cockburn's was the best known of the two, and Miss Doughty reports that they could be thoroughly relied upon for everything—from a houseboat to a meat safe or an oven, from foodstuffs to local wine (she particularly recommends their Kashmiri Medoc and Barsac).

Even the tradesmen who pestered the European visitors so mercilessly won Miss Doughty's affection, for, not only were their wares temptingly pretty, but they themselves were touchingly trusting of their British customers: 'They will take cheques on un-heard of banks, receive IOU's on scraps of old paper as if they were coin of the realm, and cheerfully consent to manufacture and despatch wares to London without pre-payment ...'.

Miss Doughty's own interest lay in the direction of botany, and although she hired a *doonga* and prettified it with hanging baskets of flowers and embroidered curtains and cushions, she spent less time on the boat than on her feet tramping through the hills, painting and collecting flower specimens in a state of wonderment and happiness. Some days it seemed to her that the valley was all blue and white like a willow pattern plate—'blue skies and white clouds, blue hills crowned with snows, blue iris shadowed by white fruit blossoms ...'. At other times she was dazzled by the brilliant colours of the wild red roses and violet irises—'The sheer gaudiness would have frightened a painter, who would never have found spectators sufficiently credulous to put faith in his portrait . . .'. And once she found herself in a world of only white flowers—'Kashmir in her spring wedding garment . . . an altogether unforgettable vision of loveliness.' Altogether Miss Doughty collected 206 species of plants during her stay in Kashmir, and these are listed in her book along with apologies for her occasional lapses into idle rapture.

Two years after Marion Doughty had stood spellbound in a Kashmir meadow full of white flowers, and Florence Parbury and her mother had strummed their guitars under the Kolahoi glacier, yet another British memsahib, Margaret Cotter Morrison, took up her pen to write the blow-by-blow account of *her* visit to the valley. At first, neither Miss Morrison's holiday, nor her description of it, which has the magnificently down-beat title *A Lonely Summer in Kashmir*, look particularly promising. Poor Margaret Morrison had not intended to be lonely at all—she had arranged to spend an exciting few weeks exploring Kashmir with a friend, but no sooner had the two met up in Srinagar when a telegram arrived summoning her companion back to India and leaving Miss Morrison high and dry 'with no friend in the land and no one with whom to travel or chum'. To begin with, she thought her best plan would be 'to hang around Srinagar or the hill-station of Gulmarg' and integrate herself into the society of the other British, but after thinking about this she sensibly decided that the British were the same wherever they were, but there was only one Kashmir—and no real reason why she should not tackle it alone.

She did not linger long in Srinagar, but hired a *doonga* and embarked on her first excursion, a fairly standard one, up-river to Anantnag, with detours (on hired horses) to examine the various sights en route such as Achabal and Martand. But Miss Morrison felt a pleasurable frisson of excitement when the *doonga* got under way and she heard its boards creak and the sound of rushing water under its keel. It is amusing to observe Morrison becoming bolder and bolder and more and more self-sufficient with each new adventure in Kashmir. On that first journey, she had set off armed with her own tea-table, wicker furniture, dining chairs, damask table cloths, rugs and a hat box, but by the end of her holiday she was travelling light and writing: 'One realises so soon that the palm of the hand makes a very good cup, the back can be used for a serviette.' She felt no alarm at being in the wilds on her own—'I never once felt that bodily fear which a woman alone is apt to feel in most civilised countries'—but she was vastly irritated by the supposed guides and interpreters and guardians who lay in wait for visitors at all the well-known sites.

At length Miss Morrison completed her tour of man-made attractions, and set up camp in the pine woods of the Lidder valley at Pahalgam. She loved it there, and Pahalgam in those early unspoilt days sounds delightful. There were two small branches of Srinagar shops to supply vital necessities, a post-box nailed to a tree, and if you were invited out to dinner you took your own plates and mugs. A missionary arrived while Miss Morrison was there and sent a messenger around to tell all the campers that there would be a Sunday service in his tent. The messenger carried the missionary's clock so that everyone could synchronize their watches for, as Miss Morrison says, 'time in different camps could be an hour different depending on the vagaries of the watches people owned.' At the appointed hour, the congregation came strolling through the trees, their servants coming along behind carrying their chairs. During Miss Morrison's time at Pahalgam this congregation was almost entirely made up of British women and children, for most of the menfolk were out in the mountains trekking, shooting or fishing. As a result, the young English girls were so bored that Miss Morrison wished she could change sex 'so that everyone could have a jollier time.'

After four weeks in camp, and a tough march back to the capital via Lidderwat and the Tarsar lake, Miss Morrison was quite excited to be in Srinagar again. She stayed with a new friend, a lady doctor who lived on the Bund, but the city was too hot to be pleasant, and soon, like the other British visitors, Miss Morrison retreated to Gulmarg.

In those days the road from Srinagar to Gulmarg stopped three miles short of the resort itself, at a place called Tangmarg. ('*Tang*' means 'narrow' in Hindi.) Here the mountain became too steep for wheeled vehicles and visitors had to complete their journey on foot or pony. 'An hour's toiling up the woodland path brought us to the entrance of Gulmarg', says Miss Morrison in what is probably the liveliest description of the place as it was nearly a hundred years ago:

on the right stands the Residency, on the left the doctor's house... straight in front, lying slightly in a hollow, are the grassy plains of Gulmarg, bearing many signs of civilisation on their two or more miles of undulating greenness. Most conspicuous is the Maharaja's palace, bare and ugly as a barrack; opposite to it on a knoll on the left, stands the little English church; eager golf-players in twos and fours can be seen following the winding course of the links from green to green; a polo-ground, club-house, tennis and badminton courts, occupy a central position accessible to all; further on the rambling buildings of the hotel stand on a spur of land all to themselves; all round are roughly built little wooden houses, barely more than huts, which is the name they generally go by, but which yet command a very fancy price during the season....

Miss Morrison found Gulmarg a pleasant little place where one could quickly get to know almost everybody and feel at home. Gulmarg was full of military men on leave from lonely, outlying districts of India, men who were determined to enjoy their precious few weeks' freedom. 'There was never a dearth of men for a picnic up the mountain to Killanmarg', writes Miss Morrison cheerfully—and there was no lack of partners at the dances held in Nedou's Hotel, nor at the dinner parties in private houses. Miss Morrison wrote that going out to dinner in Gulmarg was an experience in itself, for it meant hiring a dandy (a chair with poles back and front which rested on two bearers' shoulders), as well as a man to go ahead with a flaming torch and light the path.

The fitful glare of the torch, the curious rhythmical grunt of the natives as they picked their way bare-footed down the path, the dazzling galaxy of stars overhead, the mysterious look of the 'marg' at night time given up to solitude and a few grazing cattle... the will-o'-the-wisp twinkle of other torches guiding similar revellers to their tryst, all disposed the mind for something unexpected and romantic; and there was always a slight shock of surprise when this preface ended in a pretty drawing-room with softly-shaded lamps, ladies and gentlemen in immaculate evening dress, French menu cards on the table and a dinner served up as it would be in Mayfair.

Towards the end of August Miss Morrison knew she must leave Gulmarg and press on if she was to accomplish a trek up the Sind valley that she had set her heart on, so she ordered tents and camping equipment from Cockburn's Agency, asked them to provide a competent man to manage the coolies, and arranged to rendezvous with them in Srinagar in a week's time. Meanwhile, she flung herself into enjoying her last few days in Gulmarg.

Miss Morrison's new-found friends in Gulmarg had implored her not to undertake the Sind valley trip without a companion, and she had begun to feel nervous about it herself. But in the end the only problem with the Sind valley turned out to be that it was too *tame*—there was a telegraph wire running through it, and, Miss Morrison writes, it was 'only like make-believe at being away in the wilds, when at any moment I could drop a message into Srinagar if anything happened to me'. She continued on to Sonamarg and found that more exciting—storms were lashing the mountains as she pitched her little tent near the pathway which her guide-book told her was the high road to Central Asia, and the vastness of it all exhilarated her. This was her last trek and, being anxious to prolong it as long as possible, she made an unscheduled detour on the way back, up the Erin valley to the foot of Harmukh mountain. She was rewarded by seeing the ruins of the two Hindu temples at Wangat, as well as the Gangabal lake—dramatic and desolate, and a view of the Kashmir valley so spectacular that it was more like a map than a landscape. The night she spent on that mountain side was so cold that her sponge froze into a block of ice.

Back in Srinagar, Miss Morrison packed her souvenirs—some of them purchased from a 'charming old dealer in papier mâché known to Europeans by the alluring name of Suffering Moses', and prepared to leave Kashmir—proud and pleased that she had not failed the valley's challenge. Even the friends who had urged her not to be quite so intrepid were glad to be proved wrong, and she spent her last evening in Srinagar among them. Then,

as the light grew dim, we strolled across the *maidan* to the Club House on the embankment, where the tall cedars stand overhanging the river, and people stood in groups chatting, while the glow died away westward in the sky; small boats crowded the steps, waiting to take home their owners, people called to their servants to carry the books they had chosen from the library, the dogs got in everyone's way.... Then each stepped into his canoe, goodnights were called, and to the splash of paddles each one was speeded to his boat or bungalow, which, with bright lights and coloured lamp shades, looked very attractive from the outer darkness.

With this rather haunting description of twilight in Srinagar we can leave the British in Kashmir, for there were no great changes in the way of life they led there from the time of Miss Cotter Morrison's visit, to Indian independence. Motor cars arrived, of course, and roads improved, and fashions altered fundamentally, and a stream of weary soldiers came on leave during two World Wars—but the bridge parties at the Club continued exactly the same, and so did the picnics and dances and dinners and games of golf in Gulmarg. The same English sweet-peas and delphiniums bloomed in the garden of the Residency where, year after year, guests enthused in just the same way over the strawberries and cream served at the parties held on the smooth green lawns. And all through these years Kashmir, with its romantic reputation, remained the favourite spot for British honeymoon couples.

And then abruptly, these particular honeymoons ended. In 1947 the British left India, and in Kashmir another era came to an end, another group of strangers pulled out, leaving behind them a few relics of their culture. Not marble mosques and Mughal gardens this time, but stockbroker-Tudor houses, Hansel and Gretel wooden huts, comically christened houseboats, and tailors with names like Savilrow, who even now claim to be under the patronage of Lord Roberts, Commander-in-Chief of the Indian Army.

The annual tide of summer visitors to the valley has continued, but it has changed nationality—it is now mostly Indians playing golf or skiing in Gulmarg, and it is Indian honeymooners boating on the Dal lake. The Westerners who go to Kashmir now—and there are about 50,000 of them a year—do so on package tours and stay for only a day or two. No one—not Indians, nor British, nor French, nor Australians, have time any more to follow Arthur Neve's kind of advice and spend 'the end of May and half of June slowly moving up to Sonamarg'. Off the beaten tracks beloved by tour operators, Kashmir is probably less well-known by its visitors than it was fifty years ago.

CHAPTER FOUR

A Thousand Flowers: Arts, Crafts, People

These are our works, these works our souls display,
Behold our works when we have passed away.

—Muslim dedication

CHAPTER FOUR

A Thousand Flowers: Arts, Crafts, People

There can scarcely be a middle-class household in Europe or the United States that does not possess something from Kashmir, quite possibly without knowing it. It might be a chain-stitched felt rug in a child's room, or a tiny flower-painted papier mâché box among a teenager's treasures, or an embroidered cushion-cover, or a finely knotted carpet that passes for Persian. It might be an old shawl—too cumbersome to wear these days, so draped over a table or a bed or a sofa instead of a body. Most probably though, it is a fabric printed with that perennial favourite, a 'paisley' pattern—a scarf perhaps, or a dressing gown, or a pair of curtains.

Few people realise what a huge design debt we owe to Kashmir, for *all* those patterns that we call 'paisleys' are only developments and variations of the ones which first came to the West in the last century woven into Kashmir shawls. That distinctive motif—it has been called a teardrop, a mango, a Cypress tree blowing in the wind—only became known as a 'paisley' because popular copies of Kashmir shawls were made in the mills of Paisley town in Scotland. The ironic part of this story is that the soft wool from which the shawls were made *did* become known as 'cashmere' when, in fact, it was never produced in Kashmir at all, but imported there from Tibet and Central Asia.

Crafts from Kashmir crowd the shelves of Indian emporiums and gift shops around the world and are probably familiar to us all, and yet few of us appreciate that they have had a long and distinguished history, and that the forerunners of all the bright little boxes and bangles, the carved screens and tables, the rugs and carpets, were held in the highest regard by a most sophisticated and surprisingly international clientele for many centuries.

In 1589, when Emperor Akbar paid his first visit to Srinagar after his conquest of the valley, the court historian wrote: 'This has been a flourishing city from ancient times and the home of artificers of various kinds'. In 1611 William Finch, an English adventurer living in Lahore, noted that not far away, in the 'goodly plaine' of Kashmir 'are made the rich pomberies [shawls] which serve all the Indians.' Fifty-four years later François Bernier who, having lived in Paris and at the Mughal court presumably had a shrewd eye for quality and artistry, described the Kashmiris as skilful and industrious—'The workmanship and beauty of their palanquins, bedsteads, trunks, inkstands, boxes, spoons and various other things, are quite remarkable, and articles of Kashmiri manufacture are in use in every part of India'.

In 1783 George Forster was surprised by the numbers of foreign buyers in residence in the city of Srinagar: 'In Kashmir are seen merchants and commercial agents of most of the principal cities of northern India, also of Tartary, Persia and Turkey . . .'. It was a Georgian trader, remember, who helped George Forster escape from the country. Like Bernier, Forster was impressed by the country's craftsmen—'The Kashmirians fabricate the best writing paper of the east, which was formerly an article of extensive traffic, as were its lacquer ware and cutlery . . .'. Three years after Forster's visit, a Russian traveller, Philipp Efremov, described the Kashmiris as 'adroit artisans who also like to engage in trade'.

Kashmir's flowering in arts and crafts began with the conversion of the country to the Muslim faith. When Kashmir turned to Islam, strong links began to grow between the valley and the rest of the Muslim world, particularly Iran. A two-way traffic developed—of Kashmiris visiting religious establishments and shrines, including Mecca, in other Muslim countries, and naturally of travellers and holy men from those countries visiting the valley so famous for its beauty. In 1372, for instance, a mystic called Shah Hamadan arrived in Kashmir with hundreds of followers. As his name reveals, he came from Hamadan, a town in Iran famous for its carpets, woollen stuffs, felts, and copper utensils, and it would have been odd if during their long stay in Kashmir he and his Iranian disciples had not communicated some of their taste and knowledge on the subject of these crafts to the Kashmiris.

A little later on, when Zain-ul-Abidin, the future king of Kashmir, was growing up, a tutor was chosen for him who encouraged and strengthened all these connections between the valley and the outside world. This was Maulana Kabir, a Kashmiri who had himself been educated at Herat, an Iranian city famous for its beauty and learning and wealth of talented artisans. It was hardly surprising that when Zain-ul-Abidin came to power he made Persian an official language of the court and sent ambassadors to centres of Muslim culture, both near and far (far included Turkey, Mecca and Egypt). It was only a short step from there to inviting some of the artists and craftsmen working in those centres to come to Kashmir and teach his own people their skills.

A huge variety of craftsmen accepted his invitation. Wood-carvers came, and paper makers, and men who could shape and paint delicate objects in papier mâché (a material which was not known in Europe until the seventeenth century), and shawl-weavers, and all manner of other embroiderers and workers in textiles whose descendants still practise some of their skills. Chain-stitch embroidery remains one of Kashmir's specialities. It is coarsely done in wool on felt for rugs called *numdahs*, finer and stitched onto canvas for more elegant carpets and rugs, and finest of all on soft woollen fabrics for shawls and gowns and wraps—not forgetting crewel work, also a chain stitch, which is done in wool on cotton for cushions and curtains. Other uniquely Kashmiri embroidery work is seen in the floor coverings called *gabbas* made out of thick woollen cloth, usually old blankets, cut out and stitched one on top of another in complicated appliqué patterns. There is a nice story that the Sikh leader Ranjit Singh once ordered a *gabba* for himself. When it arrived he was so delighted that he threw himself down on the floor and rolled himself up in it.

Kashmir is also famous today for her fine, hand-knotted Persian-style carpets and rugs. This art was probably introduced into Kashmir along with all the others, but somehow declined or died out after the Mughals left the country. It was resurrected by British and French dealers in the last century with great enthusiasm, and Kashmiri carpets were exhibited even at the Chicago World Fair in 1890. Lord Curzon is said to have bought, for a mere £100, a Kashmiri-made copy of the famous Ardabil carpet, which hangs in the Victoria and Albert Museum in London, that could hardly be distinguished from the original.

Vincent Robinson, who was a well-known carpet dealer in London in the last century—in fact, he sold the original Ardabil carpet to the Victoria and Albert Museum in 1893—loaned two lovely antique Kashmiri carpets to the Museum. One was in pale yellow and the other in shades of rose pink, and they were thought to have been made in Kashmir in the early eighteenth century when the Mughals still held the country. Robinson also dealt in new carpets from Kashmir, and exhibited a spectacularly large one of these at the Paris Exhibition of 1878. This gave Sir George Birdwood (who wrote the handbook to the Indian section of that Exhibition) an excuse to contrast the new most unfavourably with the old. Robinson's new carpet was, he wrote, 'a striking illustration of the corruption of native design under European influences'. The wool of modern Srinagar carpets was 'good', the texture 'not bad', but, Birdwood continues, the 'agonized contortions' of the designs indicated that it was 'hardly possible that they can ever again be made to satisfy a critical taste'.

Another important group of craftsmen who set up shop in the valley at Zain-ul-Abidin's invitation were the artists in metal: enamellers who worked in silver or copper or brass and whose favourite colours of dark green and blue and pale turquoise and brownish-red—all made from semi-precious stones—came to be looked on as typically Kashmiri; engravers whose superb patterns were deeply grooved into silver or copper or brass by means of chisels and hammers, and sometimes dramatized by being filled in with black lacquer. Other craftsmen were specialized in the art of gilding silver, and still others at damascening iron—that is producing a pattern on a sword blade or gun barrel by a process involving corrosive liquid. Cecil Tyndale-Biscoe wrote that the coppersmiths' bazaar in Srinagar was 'easily known by the continual din', but that it was worth a visit—it is still there, not far from the Zaina Kadal bridge. 'One can sometimes pick up some really elegant and quaintly shaped jugs and basins of ancient make', he goes on , but warns that there are also 'excellent imitations of the same, which are sold to the unwary as the real article'.

Tyndale-Biscoe's own favourite piece was a cunning gadget designed to blow up a fire. This was a sort of copper jug shaped like a duck but with no opening in it anywhere except at the end of its long, downward-curving beak. It worked like this—the duck, with a little water inside it, would be placed on an ailing fire. When the water inside boiled, steam hissed out through the hole in the beak and blew on the coals, encouraging them to glow. (Sometimes the jet of steam coming out was so strong that it blew the whole fire out of the grate.) When the duck was empty but still hot, the beak was held in water which the duck sucked up into its empty belly, ready for use next time the fire looked like going out. 'This jug', says Tyndale-Biscoe, 'takes the fancy of most visitors.'

Other favourites among the early tourists were engraved brass or copper trays which were perched on wooden legs or set into table tops; versions of these were ordered by the hundred. Towards the end of the last century European visitors thought up a new use for copper ware—it became fashionable to have it finely engraved or chased in Kashmir, often in a paisley pattern, and then sent to Britain to be electroplated to look like silver.

No one knows now from where exactly each group of craftsmen moved to Kashmir, but the shawl-weavers and the papier mâché painters are presumed to have come from Iran for, like the Iranians, they were Shia Muslims. The weavers and their families were decimated by the Kashmir famine of 1877 but the papier mâché craftsmen survived, and to this day they form a tight-knit community, proud of their strong links with Iran and of their Shia faith, and fiercely protective of the secrets of their art. Many of the older generation still speak Persian and, until recently, it was the custom for them to return to Iran towards the end of their lives to die and be buried there. Down the troubled centuries the Persian connection has remained unbroken, and whenever feasible, old Kashmiris have made their last dying pilgrimages to the land of their forefathers, and young Persian artists have travelled to Kashmir to seek their fortunes. In Mughal times, particularly, there were lots of Persians in Kashmir. At least two of the governors were of Persian descent, and then again, after 1846, when the Dogra maharajas took over the valley and restored stability, there seems to have been another influx of artists. Many of today's craftsmen date the arrival of their families to that time, and the papier mâché painting done in Kashmir in the second half of the last century is very often almost identical to what was being done in Persia at the same period.

The Shias of Persia have always been less strict than other Muslims in their observance of the Islamic law forbidding the drawing or painting or sculpting of men or animals for fear of setting up false gods, and the papier mâché painters of Kashmir share this relaxed attitude. Animals and birds and men—usually hunters—sometimes appear in their work. Nonetheless, their religion is never very far away: an artist told us the sad story of his great-uncle, a well-known papier mâché painter who was working one day on a flower pattern—he had just completed a perfect rose when to his excitement and delight a butterfly came and sat on it. The painter became obsessed with how life-like his lovely rose looked, until he decided that the only thing it lacked was a scent, so he concocted a mixture of perfume and varnish and applied it to his rose so that it smelt as sweet as it looked. That night the painter was struck blind—God's punishment for daring to imitate the Creator.

The papier mâché painters and their families worship in their own Shia meeting houses called imambaras and, not surprisingly, since the painters themselves help to decorate these, there is often fine work to be found in them, particularly on the wooden ceiling panels. Sadly, however, tastes are changing. When we were in Kashmir we watched in dismay as the old painted ceiling of an imambara was demolished to be replaced by plaster-board.

Perhaps the most extraordinary thing about the craftsmen of Kashmir is that they survived at all, and were not, every one, wiped out like the shawl weavers by the continual harsh treatment, starvation wages and miserable working conditions they seem to have been subjected to after the Mughals were driven out of the valley. Godfrey Vigne, who visited Kashmir during the dark period of Sikh rule, tells a revealing story. He had commissioned a wood-carver to make him a model of a Kashmiri mosque—a souvenir to take back to England—but to his irritation the carver seemed reluctant to come and finish his work, and though he had been recommended as a skilled craftsman, when he did appear he would make ridiculous mistakes such as cutting the pillars for the mosque in different lengths. Eventually, Vigne found out what was wrong and why the poor wood-carver was so unwilling to work. It turned out that the Sikh servant whom Vigne was sending each day to fetch the carver was not only beating the unfortunate workman, but forcing him to pay him money as well. And more than that, Vigne wrote, the carver knew 'that what I paid him, or at least a great part of it, would probably be taken from him; and what was worse, that having proved himself clever enough to satisfy an Englishman that he was an adept in his profession, he would ever afterwards be made to work by the Governor for little or perhaps no pay at all.'

The miracle is that the objects made by Kashmiri craftsmen give nothing of all this away. No one could guess from the glorious patterns and rich colours of a Kashmir shawl, or from the delicious spring flowers spilling over the surface of an old papier mâché box, that the work was done in dismal surroundings by artisans who received no respect—or justice—from their employers, and had only the crudest tools at their disposal. Even today the papier mâché painters of Kashmir are less well-equipped than many a middle-class child in Europe. But Kashmiri designs are joyful, perhaps because they are nearly all floral patterns, and flowers seem to make everyone happy. One only hopes that as they etched, carved, embroidered, wove and painted their bouquets on metal and wood and cloth and paper, the artists felt some pleasure too, and that the beauty of the things they made dulled the pain of their lives a little.

Kashmiri craftsmen have never had to look far to find inspiration, for in the span of a year in the valley there are flowers enough to suit every mood—and more. First come the fragile blooms of spring: iris, narcissus, tulip, bluebell, almond blossom, crown imperial, lilac; summer brings the more flamboyant, worldly flowers: peonies, carnations and roses of every colour and size. In autumn the pointed leaves of the chenar trees turn every shade of copper and gold and red, and have inspired dozens of patterns. There are water-lilies in a surprising choice of colours—from palest pink to maroon—and, of course the exotic lotus which blooms on the Kashmir lakes in June and July, its huge flowers held high on impossibly slender stems. Then there are the birds—the kingfishers, pride of Kashmir, flashing turquoise and emerald as they dart among the reeds, the green parrot who is a summer visitor, as are the crested hoopoe and the brilliant golden oriole. And nature has inspired geometric designs too: tendrils and vines have been looped and curled and twisted into a myriad graceful arabesques, flowers and leaves have been regimented into rows, roses have been massed together to make abstract, dappled backgrounds, and even water ripples have been tamed into regularity.

As they were handed down from one generation to another, these patterns became formalized, with only small adjustments being made now and again to suit a particular customer perhaps. Families specialized in certain designs and guarded them jealously, refusing to teach them to anyone outside their immediate circle. Today a range of these patterns can be seen in the sample books kept by the papier mâché dealers and artists. In these, each brilliant page of flowers, foliage, birds and butterflies is a recognized pattern with its own Persian name: '*gul hazara*' means 'a thousand flowers'; '*gul andar gul*' means 'flower upon flower'; '*gul velayet*' means 'foreign flowers', and so on. ('Foreign flowers' is said to have got its name because it was the design that the British memsahibs liked best, but in fact it was probably a design brought from Persia—and therefore 'foreign'—by immigrant artists.)

Customers for Kashmir's arts and crafts were never lacking. Apparently even the Buddhist monasteries of Tibet regularly placed orders for a certain kind of carved and painted table used in their rituals; it was called a *skasha* and was traditionally coloured gold and red with green medallions. And when the remote Hunza tribe on Kashmir's far-flung northern frontier was subdued by the Indian Army in 1891, painted papier mâché pen cases and work boxes were found in the defeated ruler's palace.

Customers, of course, are extremely influential people, for in the end it is they who dictate the type of work done and its quality—as can be seen only too clearly in Kashmir today where some buyers among the traders and tourists, are lowering standards because they are far more concerned about cost than craftsmanship. Curiously enough, François Bernier complained about exactly the same sort of thing in 1663 when he wrote crossly that the Indians were appalling customers 'who regard not the beauty and excellence, but the cheapness of an article...'. According to Bernier, only the Mughals saved the day—'the arts in India would long ago have lost their beauty and delicacy if the monarch and principal omrahs did not keep in their pay a number of artists...'.

The Mughals' refined taste and love of beauty made them excellent clients. They demanded and got the very best from their artists, and when they conquered Kashmir their patronage gave the craftsmen of the valley a tremendous boost. On their side, the Mughals were delighted to find themselves masters of a nation of already-talented workmen. Akbar's court historian wrote appreciatively that Kashmir's craftsmen were skilled enough to be 'deservedly employed in the great cities'—and employ them the Mughals did. Manuscripts produced in the Mughal court workshops after the end of the sixteenth century often contain the names of artists with the appendage 'Kashmiri' to show that they came from Kashmir, and in the record of craftsmen employed on the Taj Mahal the name Ram Lal Kashmiri appears, suggesting that at least one artist from Kashmir had a hand in that masterpiece.

The way in which the Mughals lived gave plenty of scope for the employment of Kashmir's craftsmen—especially the papier mâché artists, who, I should explain, painted on wood just as skilfully as on objects made of paper. Bernier describes a 'good house' in Mughal times as having ceilings 'gilt and painted, but without pictures of man or beast, such representations being forbidden by the religion of the country', as well as courtyards, gardens, fountains and open terraces (where the families' beds could be placed at night to catch any breeze), and rooms with richly carpeted floors strewn with embroidered cushions and mattresses for the master of the house and his guests to relax on.

We already know from Bernier's description of it how the inside of the black marble summerhouse in the Shalimar garden in Kashmir was painted and gilt all over, and Bernier also tells us that much Mughal furniture was painted and gilded in the same way—bedsteads and palanquins, as well as elephant howdahs, tent poles and the emperor's field thrones.

A number of Kashmiri papier mâché artists may well have been employed in the workshops of the Mughal court in India, and indeed Bernier says that in the Red Fort in Delhi (where he liked to watch the various craftsmen toiling away) there were 'varnishers in lacquer work' as well as embroiderers, goldsmiths, jewellers, painters, joiners, turners, shoemakers and weavers who, incidentally, could produce muslin so fine 'as frequently wears out in one night'.

But some at least of the papier mâché work was commissioned directly from Kashmir, for when William Moorcroft was in the valley in 1823 he was told that under the Mughals the papier mâché industry had thrived and employed a large number of craftsmen who periodically sent samples of their painting down to Delhi to be inspected by the emperor. Moorcroft was shown some of these specimens—'a collection of patterns painted on plank and submitted to Emperor Aurangzeb who particularly patronised a variety of work called *subz-kar* of foliage grouped or compounded on a ground of gold and afterwards highly varnished'. This foliage pattern might easily have been one of the designs inspired by the leaves of the chenar tree which are still being painted today.

At the time of Moorcroft's visit the number of papier mâché artists working in Srinagar had dwindled to forty, and Moorcroft estimated that they produced about 1000 pen-cases a year. The demand for pen-cases never ceased in Kashmir, because traditionally the *munshis* or white-collar workers of the country—the clerks, scribes, accountants, secretaries and so on—always carried a pen-case or a scroll of paper as a sign of their trade. Indeed, the art of papier mâché painting in Kashmir is called *kari-kalamdari*, which is Persian for 'pen-case work'. According to Moorcroft, two types of pen-cases were made—portable ones with sliding lids just like old-fashioned school children's pencil cases, and a more elaborate version which Moorcroft describes as 'table furniture' and which was designed to sit grandly on a stand or a tray. Both kinds, Moorcroft says, were 'remarkable for the variety and elegance of the patterns with which they are painted, most generally of flowers, for the brilliancy of their colours, and the beauty of the varnish'.

The lavishly painted and gilt elephant howdahs and palanquins that Bernier had described being used in Aurangzeb's reign were still being made when Moorcroft was in Kashmir, and the papier mâché painters were still called upon to decorate the walls and ceilings of buildings. Moorcroft thought this a waste. He considered the papier mâché painters an 'ingenious race' and hoped that one day a wise government would direct their talents to 'loftier objects than articles of furniture or decorated pen-cases'. No wise government came along, however, and the painters stuck to what they knew. When Honoria Lawrence travelled to Kashmir in 1850, a litter was sent down from the valley to carry her which she described as 'painted and gilded like a Lord Mayor's coach'. In Rudyard Kipling's short story 'Krishna Mulvaney', published forty years later, he wrote of 'a palanquin of unchastened splendour ... rich with the painted papier mâché of Cashmere'. And the nineteenth century German art historian William Lubke particularly singled out for praise the Maharaja of Kashmir's 'handsome painted and lacquered barges and boats at Srinagar ...'.

When the Sherghari palace in Srinagar, now known as the Old Secretariat, was re-built at the end of the last century, papier mâché painters were brought in to decorate it, and, judging by the glimpses one gets of Mughal interior design in old miniature pictures, the style and quality of their work was almost identical to that being done 250 years before. The original Sherghari palace, put up by the Afghans at the end of the eighteenth century, was also painted inside apparently, for Baron Hügel visited it in 1835 and, although he was disappointed in the palace as a whole, he liked the 'japan work and carved wood'. Travellers at that time often wrongly described painted papier mâché as 'japanning' or 'lacquering', for the lacquer work of Japan and China had long been fashionable in Europe. Real lacquer is a technique involving the application of layer upon layer of the sap from the *Rhus verniciflua* or lacquer tree which dries hard and glossy, almost like plastic, and can be painted and even carved into patterns. Kashmiri 'lacquer' is merely painting on wood or papier mâché which has been varnished.

In 1911 Kashmiri woodcarvers and papier mâché painters collaborated in the old Mughal tradition to celebrate the crowning of a King-Emperor far more powerful than Akbar, Jahangir or Shah Jahan had been. In that year King George V's Coronation Durbar was held in Delhi, and the façade of the Durbar Hall was a magnificent creation of carved wood and painted panels—all made by Kashmiri craftsmen. The king admired it, and it was immediately presented to him by Kashmir's Maharaja Pratab Singh, though what happened to it after that on one seems to know.

The papier mâché painters were very far from being the only Kashmiri craftsmen that the Mughals admired and appreciated. We know from Abul Fazl, the court historian, that, of all the talented calligraphers employed by Akbar, the most skilled was a Kashmiri—'The artist who, in the shadow of the throne of His Majesty, has become a master of calligraphy is Mohammad Hussain of Kashmir. He has been honoured with the title of Zarrinqualam, the gold pen'. It seems appropriate that, according to Abul Fazl, indelible ink was invented in Kashmir.

Nowadays it probably strikes us as curious that the skill of writing—that is to say of forming letters, not composing sentences—was valued so highly by Akbar and his contemporaries that they considered calligraphy a higher art than painting. Abul Fazl explains: 'the letter gives wisdom to those that are near and far. If it were not for the letter, the spoken word would soon die, and no keepsake would be left to us of those that are gone by'. This is not to say that Akbar neglected his artists; he delighted in their skill and considered painting informative as well as pleasing and was proud of the fact that under his patronage the court artists were better supplied with the tools of their trade, and had become more proficient. The works of the court painters were laid out each week for the emperor's inspection, and if he was pleased with what he saw he rewarded the artists or increased their monthly salaries. His son, Jahangir, was even more passionate about painting and prided himself on being so knowledgeable that 'If any other person has put in the eye and eyebrow of a face, I can perceive whose work the original face is, and who has painted the eye and eyebrows'.

But it was not, in the end, papier máché work, nor calligraphy, nor painting that the Mughals considered the apogee and the triumph of Kashmiri craftsmen—it was the shawls they made. Many skills came together to produce the famous shawls of

Kashmir. Spinners created woollen yarns as fine as silk which dyers coloured with unerring taste and subtlety; pattern drawers invented exquisite, intricate designs which the weavers somehow managed to reproduce on their looms. The results of all their united labours were large, decorative rich-looking shawls which were light as gossamer and yet as warm as blankets, miraculously soft, but with enough 'body' to drape and hang elegantly.

In India in those days shawls were worn by men, usually slung casually around their shoulders with one end hanging down at the back and one in the front, and not surprisingly the handsome shawls of Kashmir were prized above all others. The film *Shatranj ke Khilari* (The Chess Players) made by Satyajit Ray some years ago showed exactly how elegant they must have looked—all the actors playing nineteenth century Indian noblemen wore genuine antique Kashmir shawls borrowed from private collections.

The first specific reference to Kashmiri shawls is by Abul Fazl, who, describing the industries of Kashmir, says: 'Woollen fabrics are made in high perfection, especially shawls which are sent as valuable gifts to every clime.' Akbar encouraged the shawl industry 'in every possible way', both in Kashmir and in I ahore where some Kashmiri weavers had set up in business, and he himself owned hundreds of shawls which were stored in the royal wardrobes according to their colours. Apparently it had been the fashion to fold the shawls into four, lengthways, before putting them on, but this had recently died out and, Abul Fazl tells us, 'nowadays they are generally worn without folds and merely thrown over the shoulder'. Akbar, it seems, liked to wear two stitched together so the wrong sides could not be seen, and this, reports the historian, looked 'very well'. By Jahangir's time Kashmiri shawls were obviously so familiar to everyone that he does not even bother to describe them but simply writes: 'The shawls of Kashmir to which my father gave the name *parn-narm*, are very famous: there is no need to praise them . . .'.

Sadly for us, shawls were not made—or expected—to survive

for centuries, and the earliest examples left are a few fragments dating from the beginning of the eighteenth century. From these, and from glimpses of the shawls in Mughal paintings, we know that the patterns of the shawls at that time were restricted to the border, and that they were usually composed of delicate sprays of flowers, exquisitely finely woven.

Sebastien Manrique, a monk from Portugal who travelled in India and visited Shah Jahan's court in 1640, was clearly impressed by the beauty of these shawls: 'These choice cloths are of white colour when they leave the loom, but are afterwards dyed any hue desired and are ornamented with various coloured flowers and other kinds of decoration, which make them very gay and showy'. According to Manrique's rather charming description, the princes and nobles of the court then wore their shawls like cloaks, 'muffling themselves up in them, or else carrying them under their arms.' About twenty-five years later, in 1665, François Bernier described the size of an average shawl as being about five feet long, two and a half feet wide, and with decorated borders less than one foot deep.

Even when their patterns were limited to narrow borders, the shawls were difficult and time-consuming to make and their prices were high. According to Abul Fazl, a shawl in Akbar's time could cost anything from Rs 200 to Rs 1200, and they were always greatly valued as presents—and of course, as bribes. Sir Thomas Roe, ambassador of the English King James I to Jahangir's court, wrote how he was pressed to accept a 'Gold Shalh' by the Governor of Surat soon after his arrival in India. But he refused, not wishing to be under any obligation. Nadir Shah, the Persian leader, apparently considered Kashmir shawls quite prestigious and valuable enough to be included in the fifteen elephant-loads of lavish gifts that he presented to the Sultan of Constantinople in 1739. He may very well have acquired the shawls that he gave to the sultan when his army sacked Delhi earlier that same year. And when the British sold Kashmir to Maharaja Gulab Singh, they wrote into their agreement that every year he must present them with three pairs of Kashmir shawls.

In 1848, a year or so after the signing of this treaty, a wealth of Kashmir shawls—a vast hoard of them—fell into British hands when the Sikhs were defeated for the second time, and their capital, Lahore, with its citadel containing the *toshkhana* or treasury was taken over.

Those permitted to see the Sikh *toshkhana* before its treasures were dispersed (many of them went to the East India Company's collection in London) were bowled over. Colonel Robert Adams described it in a letter home to his cousin Lady Login:

> I wish you could walk through the same Toshkhana and *see its wonders*: the vast quantities of gold and silver: the jewels not to be valued, so many and so rich: the Koh-i-noor, far beyond what I had imagined; Ranjit's golden chair of State; silver pavilion . . . and, perhaps, above all, the immense collection of magnificent Cashmere shawls, rooms full of them, laid out on shelves and heaped up in bales—it is not to be described!

It might seem odd that Colonel Adams should have passed over the Koh-i-noor diamond so quickly and described the shawls so enthusiastically, and the explanation is this—he was writing to a lady, and by that date, a rather extraordinary thing had happened: a positive passion for Kashmir shawls had gripped the women of Europe, who now looked upon this eastern male garment as the height of western feminine chic. To own a Kashmir shawl had become every woman's dream. No one knows exactly what sparked off this craze, but one explanation is that Napoleon came across the shawls in Cairo when he conquered Egypt in 1798 and brought some back as a gift for his new wife Empress Josephine, who loved them and launched them into fashion. At one time she is supposed to have owned 200 shawls. Certainly, the smart ladies of Paris were wearing Kashmir shawls soon after this time, for they appear in several portraits painted by Ingres at the beginning of the 1800s. However, there are mentions of a shawl—and presumably one from India—much earlier than that, in 1767, in letters from the English novelist Laurence Sterne to his lady love, Eliza. She was the wife of an official of the East India Company, and he had met her when she came to England to put her children into school. When she returned to Bombay he wrote passionate letters to her: 'I kiss your Picture—your Shawl—and every trinket I exchanged with You . . . I dreamt . . . that thou camest into the room, with a shaul in thy hand . . . you folded the shaul about my waist, and, kneeling, supplicated my attention.'

In the east, the trade in Kashmir shawls was very extensive and it would have been strange if some of them had not found their way to England and France with returning French officials or East India Company employees and their wives and daughters. And then it is not difficult to see why European women fell in love with them at first sight; for, quite apart from the fact that they were astonishingly beautiful, the shawls fitted in with the strong feeling for all things eastern and exotic that was current towards the end of the eighteenth century, and perhaps most important of all, they were terribly practical: shawls could be worn elegantly over any shape of dress and, though light, they were warm as toast—altogether perfect for Europe's uncertain climate. For all these good reasons, Kashmirs or Kashmeers or Cashmeers or Cashmeres, as they became known, remained in fashion for no less than a hundred years, and they appear in dozens of pictures of women over that period, from Ingres's grand society portraits to humble advertising posters. Queen Victoria herself was fond of them, though none remain in the royal collections today—perhaps because she liked to give them away as wedding presents.

Even before the passion for shawls reached its height, weavers in Britain and France were competing fiercely with each other to produce the most successful copies of the original Kashmir shawls. In 1812 a breakthrough in weaving techniques put Paisley in Scotland well ahead in the race with authentic-looking versions of the shawls selling for about £12, whereas a genuine Kashmir shawl cost from £70 upwards. To inspire fresh designs, new shawls from Kashmir were rushed to Scotland the moment they arrived from India, and Paisley held its own for a few years. But then a loom was invented by a Frenchman called Joseph Jacquard at Lyons which could produce extremely complicated designs semi-automatically, the pattern being dictated by a punched card. Poor M. Jacquard was nearly lynched by French weavers who could see themselves made redundant by his machine, but in the end the Jacquard loom was accepted and put France firmly ahead in the shawl business.

With all this rivalry between the British and French, each trying to attract customers away from the other, it is not surprising that the patterns on the shawls became increasingly elaborate and flamboyant and soon spread far beyond the original narrow borders to cover the whole cloth. The 'paisley' motif itself—more accurately called a teardrop or cone—kept changing too, growing more and more elongated until it swirled all over the shawl. The origins of this motif, incidentally, have been the subject of much speculation. Some say that it was inspired by the great loop that the river Jhelum makes near Srinagar, others that it is based on the shape of an almond or a Cypress tree blowing in the wind, or that it was copied from the jewel that eastern potentates wore in the front of their turbans, or even that it came from the shape of the side of a person's hand when the fist is clenched. A more exotic theory is that it is the male half of the Chinese yin/yang symbol, though the female half of that is paisley-shaped too. Yet another, rather opposite view is that it represents the external female organs. ('In which case', said a woman who heard this theory, 'why don't we all walk with a limp?') John Irwin who was Keeper of the Indian Section of the Victoria and Albert Museum, put forward a less far-fetched idea in his book *The Kashmir Shawl* (1973). He believed that the motif started life as the uncomplicated flowering plant that decorated the early shawls, and that this design simply became more and more stylized as the years went by and fashion's pressure for novelty increased, until it evolved into what we now know as the 'paisley'.

By the middle of the nineteenth century a topsy-turvy situation had developed in the shawl trade: copies of Kashmir shawls made in Britain and France were being exported as far as the Middle East, while in Kashmir itself French agents were persuading Kashmiri weavers to copy 'eastern' patterns designed in Paris. An Englishman won first prize for his Kashmir shawl design at an exhibition in Lahore in 1873, and Jacquard weavers in Europe were 'signing' their shawls with fake Persian lettering. Persia was producing its own copies of Kashmir shawls, and to make things even more complicated a Turkish trader from Constantinople had introduced a new kind of shawl that was *embroidered* all over to look as though it was woven—for it took far less time to embroider a complicated pattern onto a shawl than to weave it into the cloth on a loom.

No wonder visitors to Kashmir itself sometimes became confused. Dr Adams of the 22nd Regiment on leave in Kashmir in 1867 confessed himself baffled and disappointed, for though he saw one magnificent shawl being made there for the French empress Eugenie, 'It became clear that the numbers to be seen in London and Paris could never have been made in the shops of Serinugger—not even in a century . . .'.

A little book was brought out in England in 1875 to celebrate the visit that the Prince of Wales was to make to Kashmir's Maharaja Ranbir Singh the following year, and to educate the public on the now extremely muddled subject of the Kashmir shawl. *Kashmeer and its Shawls* was written by an anonymous gentleman who, in an effort to make his book more readable, wrote it as a series of questions and answers between a fictitious mother called Lady Ann and her Daughter Lily. Daughter Lily starts the book off by asking 'Mamma dear, I have heard a good deal about Kashmeer Shawls, and I think I have caught a glimpse of one, worn by yourself when going out to parties. It looked very beautiful, and I should like much to know what a Kashmeer, as distinguished from any other shawl is, if you please to tell me.'

Lady Ann was not reluctant to oblige and her answer takes up most of the next sixty pages. It is clear that by this time the fashion for the shawls was in decline, for she ends her lecture with the ardent hope that fashion 'will soon raise the Kashmeer shawl to its former position, and that a garment, as modest as it is becoming, putting aside its beauty, durability, and excellence, may soon re-assert, under the patronage of our much-loved

Queen and her daughter the Princess of Wales, that place in a lady's wardrobe, and on her person it so eminently deserves…'.

In the mean time, the fashion for Kashmir shawls had given a great boost to the valley's other arts and crafts, and at all the prestigious exhibitions held in Britain in the last century Kashmir was well represented—not only by her shawls, but by papier mâché, metalware, wood carving and embroidery too. The most stunning exhibits of all were sent by Maharaja Gulab Singh to the Great Exhibition of 1851. The maharaja excelled himself. Apart from a collection of fabulous woven 'Kashmirs', he sent a four-poster bed entirely made of enamelled silver with the finest shawls as hangings, as well as an enamelled silver service, wood carvings and a beautifully painted papier mâché pen-case and tray. The silver bed, described even by blasé officials as 'splendid' and 'remarkable', can be seen in the background of the official painting of the Indian section in Dickinson's Pictures of the Great Exhibition.

In June the following year, there was an auction sale of some of the objects that the East India Company had shown at the Great Exhibition. The sale took place at an auction market opposite the Bank of England and lasted for two weeks. The catalogue lists dozens of mouth-watering shawls from Kashmir and the Punjab—where the expatriate weavers that Akbar had encouraged were still going strong—as well as 'a very beautiful papier mâché writing box and stand richly painted in green, crimson and gold and bouquets of flowers'.

What is now called the Victoria and Albert Museum in London was also born that year, when the government voted £5000 to purchase works of art, many of them Indian, from the Great Exhibition, so that they could be a perpetual source of inspiration to the British public. Queen Victoria loaned Marlborough House as a temporary home for this collection while a proper museum was being built for it in South Kensington, and she also presented it with some exhibits, including nine Kashmiri painted boxes.

The museum was moved out of Marlborough House and into its permanent home in South Kensington in 1857, and the first-ever catalogue of all the 'Objects of Indian art exhibited in the South Kensington Museum' was compiled by Lt Henry Hardy Cole in 1874. Kashmir was well represented in it, and Lt Cole quoted Sir Digby Wyatt, an expert on Indian art, on the subject of Kashmiri papier mâché. The painters, wrote Sir Digby floridly, possessed 'a skill rivalling that of Magna Graecia or Etruria', and 'the painting on papier mâché produced in the valley of Kashmir, is as celebrated almost as the shawls which are made there'.

In the mean time, the hoard of treasures acquired by the East India Company in the course of its long period of rule in India had also been put on view to the public in another museum, the India Museum, which opened in Whitehall in 1861. More than 175,000 visitors passed through the gates of the India Museum in the first two years, all eager to see its fabled curiosities and wonders, which included Ranjit Singh's golden throne and other riches from the Sikh *toshkhana*. In 1880 the India Museum amalgamated with the South Kensington Museum. Together they possessed a unique and fabulous collection of arts and manufactures from India with, naturally, Kashmiri shawls, papier mâché, embroideries, metal work and wood carving among them.

Between 1880 and 1882 the Museum's collections were boosted still further when Caspar Purdon Clarke, the authority on Indian art, who later became Curator of the Museum, was sent to India on a buying trip. Sir Caspar collected 3400 items, including many from Kashmir. The money for these purchases, incidentally, came from an exhibition of the presents given to the Prince of Wales on his visit to India a few years before. In 1889 the South Kensington Museum changed its name, by order of the queen, to the Victoria and Albert Museum.

Today, sadly, lack of space dictates that only a small proportion of the treasures in this vast store-house—the nation's attic—can be displayed to the public, and there is usually very little from Kashmir exhibited, if anything at all, apart from shawls. The only consolation for this is that at least the papier mâché pieces kept in the crypt remain in pristine condition. They cannot fade for they rarely see the light of day.

Exhibitions were the passion of the Victorian age with its zeal for commerce, industry and trade. They came thick and fast upon each other's heels, and at all the relevant ones Kashmiri arts and crafts were on display. The major ones were the International Exhibition of 1862, the Colonial and Indian Exhibition of 1886, the Glasgow International Exhibition two years later, and the Empire of India Exhibition of 1895. Unfailingly decorative and pretty, Kashmiri arts and crafts were always popular at such exhibitions. Indeed, after the British Empire Exhibition in 1924 an official report stated that 'carved wood articles, papier mâchés and embroideries were the articles most in demand', though the organisers were quite surprised by the number of customers there had been for curry powder and incense sticks.

Meanwhile, across the Channel at the Paris Universal Exhibition of 1878, there was an Indian Court in which many items from Kashmir were exhibited—including the huge carpet which Sir George Birdwood so disliked. Sir George, however, was even more concerned and worried about the state of the shawl-weaving industry in Kashmir than he was about their carpets, for by that time some French dealers were designing shawl patterns in France and only sending them to Kashmir to be worked. 'The Cashmere shawl trade is of the highest antiquity and importance', wrote Sir George 'and it is very deplorable that it should have been recently checked owing to the use of French designs . . .'.

The tragic part of this whole story is that while experts like Sir George wrangled about the introduction of French patterns and dyes, and while the wealthy women of Europe rushed to view the last imports from Kashmir, and while merchants and shawl dealers grew fat on the proceeds, the shawl weavers themselves lived in a state of piteous semi-slavery, and their condition worsened with each new conquest of the valley.

According to the traveller George Forster, the number of shawl looms in Kashmir had dwindled from 40,000 in Mughal times to only 16,000 after the Afghans invaded because of the 'heavy oppressions of the government'. Forster did not particularly like the Kashmiris, but he did admire their skill at making shawls and other beautiful things and says: 'the quality of these manufacturers clearly evince, that were the inhabitants governed by wise and liberal princes, there are few attainments of art which they could not acquire.' However, the next conquerors of Kashmir, the Sikhs, proved to be even less wise and liberal than the Afghans. In 1843 Erich von Schonberg described the 'miserable condition' of the shawl weavers who earned about four annas a day, of which two were repaid to the government straightaway in tax, while the remaining two had to be spent on food which the government forced its employees to buy from its own stores rather than more cheaply on the open market. The Sikh system, explains von Schonberg,

> is never to allow the workman ready money, the government provides clothes, firing, and other household necessaries, charging as usual, a hundred per cent profit. This is managed very skilfully, and so arranged that the poor artisan is always in debt; and I will add, that the shawl weavers seem to be the most unfortunate.

The childhood of the weavers' children ceased abruptly at about the age of five, when they were considered old enough to work on the looms and contribute to the family's meagre budget—'and thus another human being enters on a career of wretchedness, and rears children, who, in turn, become heirs to his misery', von Schonberg says sadly.

Later, under Maharaja Gulab Singh their lives became even more intolerable: shawl weavers were required to pay still more tax and a new law was introduced forbidding any weaver, whether ill or half-blind or old and tired, to abandon his loom unless he could find someone to replace him. Though it was also forbidden to leave the valley without the maharaja's permission, many weavers risked their lives to escape from the valley at this time, and those who succeeded joined other expatriate weavers in the Punjab. But the shawls from the Punjab, although greatly admired, were never considered to be in quite the same league as those from Kashmir.

It is no accident that the most detailed description of how Kashmir shawls were made has come to us from William Moorcroft, for he went to Kashmir in 1823 with the express intention of finding out everything he could, and passing the information on to British weavers so that they could steal a march on their French competitors. He had already tried to help the British shawl industry on a previous expedition to Tibet in 1812, when he collected fifty shawl goats to send to England for breeding. This scheme failed, because the male goats were loaded onto one ship for the voyage home and the females onto another—and the boat carrying the females sank.

The very first step in the making of a shawl was the arrival of the shawl wool in Kashmir. The famous soft wool that to this day bears the name Cashmere did not come from the valley at all: it was the soft underfleece of a breed of mountain goat found far away in Tibet and Central Asia. The finest quality came from wild goats and was collected bit by bit in summer when they moulted, but excellent wool came from the domesticated herds of these goats as well. The fleeces were tightly packed in bales and came down to Kashmir on horseback, via Leh in Ladakh—a long and hazardous journey through the mountains. Once in Kashmir they were sorted, cleaned, graded and spun. There were 100,000 women and girls involved in spinning in Moorcroft's day, and they worked, he tells us, 'with little interruption' from daybreak until after dark at night, if they could afford oil for their lamps.

A great many skilled people were involved in the making of a single shawl: First the pattern-drawer who planned the design, then the specialist who chose the colours and estimated how much yarn should be dyed in each of the shades, and then the dyer who could produce as many as sixty-four different tints using natural ingredients—indigo for blues and greens, logwood for red, saffron for orange and yellow, and so on. Moorcroft tells us that a few colours were obtained by boiling down European cloths to extract the dyes. The dyed wool now went to yet another expert who cut it into the correct lengths required for the weft and the warp, and then a loom-dresser threaded up the loom. In the mean time the designer and colourist had translated their pattern into row-by-row instructions that could be called out for the weavers to follow. If the design covered a large part of the shawl and not just narrow borders, several looms would be involved, each weaving a small piece of the pattern, and at the end all the bits would be sewn together so expertly that the seams were almost invisible. It took a long time to make a shawl. Godfrey Vigne estimated that even with six or seven looms at work, each manned by two weavers, it would take six months to make a pair of 'very large and handsome shawls'.

When, at long last, a shawl was complete, it was handed over to a finisher who trimmed off straggly tail ends of wool, tidied up the back of the cloth and corrected any mistakes by embroidering over them with a needle.

After this the shawl was taxed and stamped, and then it was partially washed before being shown to the merchants who inspected each one very carefully, looking for holes and imperfections. If any flaws were found, the shawl went back to the finisher, but if it passed it would be washed once again, stretched out to dry and then packaged up for export. This was done by wrapping the shawls in greased paper and then sewing them into raw animal skins. These shrank when they dried, compressing all the shawls together and greatly reducing the size of the parcel.

At the time of Moorcroft's stay in Kashmir there were buyers and merchants from Chinese Turkestan, Uzbek, Tartary, Kabul, Persia and Turkey, all living in Kashmir. Shawl patterns were designed with each of their particular tastes and markets in mind. Moorcroft collected thirty-four different shawl patterns and sent them home to England as examples of what was being made and for whom, but sadly only eight of these have survived—three rather florid designs for the Russian market, two simple ones for India and three for Persia. They are all in the Metropolitan Museum in New York.

The dealers in Srinagar did their business through shawl-brokers, and these brokers, Godfrey Vigne tells us, competed ferociously with each other for new customers. Some of them kept informants in the larger cities of India whose job it was to alert them when a likely buyer was on his way to Kashmir. Then the broker would quickly send a messenger to invite the merchant to stay with him whilst in Srinagar, or he might send someone to accompany the merchant on his journey to the valley, and make sure that no other brokers got their hands on him. At the very least, the merchant would be met by a solicitous guide somewhere en route, as Vigne describes:

> when the merchant, half dead with fatigue and cold, stands at length on the snowy summit of the Pir Panjal, or either of the other mountain passes, he is suddenly amazed by finding there a servant of the broker, who has kindled a fire ready for his reception, hands him a hot cup of tea, and a kabab, a delicious kaliaum, and a note containing a fresh and still more pressing invitation from his master. Such well-timed civility is irresistible.

Traditionally, the shawl brokers and the dealers lived in the large wooden houses that still line the banks of the Jhelum river in Srinagar, and that is where all the business of buying and selling was transacted. Late afternoon was the favourite time for displaying the Kashmirs that were for sale, for the golden light at that moment of the day suited the shawls to perfection, enhancing and enriching their already superb colours and making them appear irresistible.

The shawl weavers themselves, however, lived and worked in dingy hovels in the squalid back streets of Srinagar, where from time to time European visitors would wander, usually by mistake, and be shocked at the poverty they saw around them. Two medical men from Britain, Dr Adams and Dr Wakefield, on completely separate visits to Kashmir, both commented on the pallid faces, stunted bodies and generally wretched and unhealthy appearance of the shawl weavers, even compared to other Kashmiris. Neither of these doctors would have been the least bit surprised to learn that in the appalling famine that swept the Kashmir valley in 1877, not long after their respective visits, the shawl weavers were decimated; they had no reserves to fall back on, no health, no strength, no money; the famine almost literally wiped them out. And when the weavers died, their art died with them. The splendidly-patterned, woven woollen shawls were never again made in Kashmir, for soon no one remembered how. The shawl industry, that goose that had laid golden eggs for Kashmir since Zain-ul-Abidin's time, and survived decades of bullying and ill-ṭreatment, had finally been killed off by greed and stupidity, and the few shawl-weavers that were left were quickly absorbed into the expanding carpet business.

No one even mourned this loss, for in Europe the vogue for shawls was over and the demand for them had come to a full stop. Nothing kills a fashion so effectively as over-popularity, and in the second half of the nineteenth century Kashmiri shawls, or copies of them, were everywhere—Paisleys cost only a pound or so by then, and shawl patterns were being printed onto cottons selling for only a few shillings. The Kashmir shawl, once a symbol of all that was rare and special and exotic and beautiful, was now considered commonplace, even vulgar.

As for the shawl-weavers of Kashmir, there was to be no justice for them, even in death. For what is the name that clung to the wonderful designs they left us? Not that of Kashmir, but of a small town in Scotland called Paisley.

Perhaps because of its past history, the whole business of Kashmir shawls today seems to be surrounded by mystique—or at least some confusion. The beautiful old shawls discussed in this chapter can no longer be bought new in Kashmir. Attempts are being made to revive the industry, but nothing has yet been produced that can compare with the shawls being made a hundred years ago. The old woven shawls *can* be found—at a price—in antique shops in both Europe and India (where they are called *jamawars*), but very often what you are being offered is only a slice off an old shawl—for the originals were not often conveniently sized, but were vast pieces of cloth that were worn doubled or folded for warmth in a way that we would find impractical today.

The most prized shawls from Kashmir nowadays are *shahtush* (meaning king's wool) shawls. These bear absolutely no relation to the antique shawls and are not very exciting to look at; indeed, to an untrained Western eye they seem rather drab. But *shahtush* shawls are unbelievably soft and light and yet extraordinarily warm. They are sometimes called ring shawls because dealers like to demonstrate how fine they are by pulling them through a wedding ring. A top quality *shahtush* can cost hundreds, perhaps thousands, of dollars, and they are status-symbols in northern India, where it is cold in winter—'Our mink and sables' is how an Indian friend put it to me. The wool they are made from does not come from Kashmir at all, but from Tibet and Ladakh and the mountains around, and it is taken from the underfleece that mountain goats grow in winter. These shawls are usually preferred in the pale brown or cream of the natural wool, but of course they come dyed in colours as well.

The next best quality shawl is *pashmina*. This is wool from the same goat, but it is of a slightly coarser grade, though *pashmina* shawls are still extremely soft and warm. Then there are various mixtures and blends of *shahtush*, *pashmina* and ordinary wool, and finally there are shawls made entirely in ordinary wool. Any of these different qualities can be bought plain, or with embroidery, or the shawl might have an exquisite border cut from an antique shawl and stitched on.

Visits to craftsmens' workshops and dealers' showrooms have always figured importantly on the itinerary of European travellers in Kashmir, and of course they still do. 'We sallied out immediately after breakfast to explore the land part of this Eastern Venice', wrote a Captain Knight in 1863.

We walked... through the most filthy and odoriferous bazaar I ever met with, till we reached the residence of Saifula Baba, the great shawl merchant of Sirinugger. Here we found a noted shawl fancier inspecting the stock, and were inducted to the mysteries of the different fabrics.... Mr Saifula Baba handed us tea and sweetmeats, after the fashion of his country; and we adjourned to the abode of a worker in papier mâché, where we underwent a second edition of tea and sweetmeats, and inspected a number of curiosities. The chief and only beauty of the work was in the strangeness of the design; and some of the shawl patterns, reproduced on boxes, etc. were pretty in their way, but as manufacturers of papier mâché simply, the Cashmeeries were a long way behind the age.

After the disastrous famine was over, visits like this continued unchanged—except that the shawl dealers could now trade only in embroidered shawls, or plain woollen ones. The other craftsmen—the metal workers, the wood carvers and the papier mâché painters—had all survived the bad years, for their work did not depend so perilously on the whims of fashion, and neither were they ever enslaved to quite the same degree as the poor shawl weavers, forbidden ever to leave their looms.

Marion Doughty, the humorous and resilient British traveller we met in the last chapter, did the obligatory tour of workshops and showrooms in 1901. Her advice to other European travellers looking for bargains was that they should try not to interfere with the local craftsmens' designs, try not to impose their own ideas or taste on him. 'He will never make a mistake when unhampered by restrictions', she wrote, and declared that the souvenirs she had taken home herself looked 'as lovely and perfect in their new surroundings as when worked under other skies for other purposes'.

Miss Doughty was particularly thrilled with the results of orders she had placed with the cloth merchant: two chain-stitched felt rugs had turned out beautifully, one in an intricate pattern of chenar leaves worked in two shades of blue, the other a design of water plants in blues and greens with red-brown stems. Though they measured six feet by four feet they cost only Rs 12 the pair. The embroidered table-cloth she had commissioned was delightful, as were the crewel-worked cushion covers which cost only eight annas (Rs 0.50) a piece, and the curtains with crewel-embroidered borders of green leaves.

Shortly after all these were delivered to her houseboat Miss

Doughty was collected by a carpet merchant who had promised to take her down-river to see his factory. There, in large sheds, she saw hundreds of workers knotting carpets according to instructions being droned out by the head boy of each loom. On the way back to her boat she was waylaid by an enamel worker who talked her into visiting his shop. She described him as 'a cunning comprehender of female character', for he insisted on giving her a cup of tea which, of course, made her feel she must buy something from him, but she was greatly impressed by the enamels he showed her, particularly some fine pieces ordered by the maharaja and destined for exhibitions in London and Paris:

They are made from crushed stones, and applied with a delicacy of fancy and richness of effect difficult to imagine unless seen. The blues made from lapis and turquoise are handsome, and so is the carbuncle red, but the beautiful dark greens and yellows obtained from agates and amber are more striking and uncommon.

These craftsmen were lucky, she felt, because their work was too slow and too expensive to tempt the ordinary tourist. Their customers were people of real taste and discrimination, and as a result they had not lapsed into short cuts and shoddy workmanship, but stuck to their traditional designs. These had been added to, wrote Miss Doughty, 'by such thorough artists as Mr Kipling of the Lahore School of Art'—this being John Lockwood Kipling, Rudyard Kipling's father, who was both Principal of the Mayo School of Art and Curator of the Central Museum in Lahore.

No sooner had Miss Doughty returned to her houseboat after this excursion than a silver merchant arrived and persuaded her to accompany him to his workshop. Another boat ride took them to an old house on the river full of stone stairways and secret courtyards, and there they found men making copper untensils:

The scene was worthy of a Rembrandt, the furnace casting a rich glow on the roundly-moulded, olive-tinted limbs of the workers, all intent on their tasks. Their tools were of the simplest—a few small wedges and punches, some nails, small hammers—and with this tiny apparatus they were turning out great standard lamps four and five feet high, trays two feet across, jugs, lanterns, all covered with the finest of traceries and reliefs.

When she enquired how Kashmiri copper acquired its particularly pretty colour, the merchant told her that it was boiled in apricot juice and salt water. 'So good housewives, take note!' she says enthusiastically, adding practically that if dried apricots could not be found, lemon juice would do as well. Miss Doughty was less impressed by some of the silver work. She was shown a coffee service designed by an Englishman that was so hideous she felt it must have been the product of a 'diseased imagination'. More tea was drunk at the silver merchant's and she was able to repay her host by translating some letters from English customers. One of these was an order from a large store for hundreds of pieces of metal ware, another was from 'a very highly placed lady' ordering a silver dressing table set, and a third was from a young army officer looking for a wedding present 'neither too costly or too rubbishy'.

The tireless sightseer now moved on to visit a papier mâché painter nearby, one of the few artists, she tells uş, who had not been corrupted by the tourist's demand for quick and inexpensive work. Miss Doughty's painter worked slowly and painstakingly and each object he produced, 'whether tiny stamp-box or larger wares, such as blotting-book or card table', was a masterpiece. 'Many of his patterns were of unknown antiquity—the 'flower patterns' of Persian origin showing a network of blossoms on a golden ground, the 'devil pattern', from mysterious Kabul, with a thousand fiendish figures mixed in inextricable confusion'. The old papier mâché painter persuaded Miss Doughty to make one final call—on a neighbouring merchant who dealt in textiles from all over the world. There was a huge crowd of people in his warehouse and she was astonished at the sumptuous fabrics for sale: gorgeous brocades, shimmering cloths of gold, brilliant silks, rippling satins and exquisite embroideries.

Modern tourists in Srinagar can enjoy.a day remarkably similar to Miss Doughty's, though it is just a little more difficult now to find the workshops of genuine craftsmen rather than 'factories' where the old skills have been sacrificed to speed and cost. In some of these, for instance, a 'hand-painted' papier mâché box is literally painted with the hand—flower petals are daubed on with fingertips dipped in colour and a brush is only used at the end to add a few unconvincing squiggles. Touts from these factories are very persuasive, and to avoid being disappointed the most sensible plan is to seek advice about whom to visit from a hotel manager or houseboat owner, or to consult one of the merchants who employ the best craftsmen—Suffering Moses, Asia Crafts, Ganemede, Qasim or Jaffer Ali. For, if you can find them, there are still carvers in Kashmir who can transform wood into lace before your eyes, carpet manufacturers who knot rugs that feel as fine as velvet, *gabba* makers (in Anantnag) whose curious cut-out and appliquéd floor coverings are unique, *numdah* embroiderers who can turn a piece of felt into a flower garden, and, of course, papier mâché artists whose delicate painted blossoms must be looked at through a magnifying glass to be truly appreciated.

In 1916 there were 150 people involved in making papier mâché in Kashmir. Today there are reckoned to be well over 1000 families involved. Of these, however, only a few are so skilled at their work that they merit the traditional title '*ustad*', which means 'master' in Urdu. This title is not restricted to the painters alone. The men who make the papier mâché shapes can—indeed should—be master craftsmen as well.

The papier mâché is, of course, made of paper. School exercise books, newspapers, whatever comes to hand, is shredded, soaked in water, pulped with a pestle and mortar, squeezed out and then mixed with paste. The sticky grey porridge that results is smoothed over wooden moulds, which are first covered with plain paper to prevent them sticking, and left to dry. This can take days and days in a wet Kashmiri spring or autumn. The papier mâché shapes are then tapped off the moulds, trimmed, filed smooth and, if they are things like vases that have to be made in two halves, stuck together with strips of paper, before being sent off to be painted. The painters work in poor surroundings with the most primitive tools. Broken cups and saucers contain their powder paints and varnishes, and brushes are locally made out of cats' hair. But the atmosphere in a serious studio or *karkhana* is almost religious in the stillness, peace and silence that radiates from the calm concentration of the artists and their young apprentices. Each painter has his own repertoire of patterns that he specializes in and teaches to the next generation. Some prefer to paint flowers, for instance, some the more difficult geometrical designs. The gold patterns are the most exciting to watch being created. They are painted onto dark backgrounds with colourless glue so that nothing much can be seen at all until the artist pats gold leaf over the apparently blank surface. Then the glue picks up the gold and a pattern emerges as thrillingly as it does in those 'magic' books for children where you rub an empty page with a pencil and a drawing appears.

Because papier mâché work has traditionally been an anonymous art with no dates or signatures on the pieces, and because the same traditional patterns and shapes were produced over and over again by succeeding generations of craftsmen and artists, it is difficult to date papier mâché accurately. Most of the pieces owned by the Victoria and Albert Museum in London were bought new, from exhibitions in Europe or directly from dealers in Kashmir, so we *do* know when they were painted, but many others remain a mystery.

A certain amount can be told from *what* an object is. Maharaja Gulab Singh is unlikely to have ordered a dainty knitting needle case, for instance, and a memsahib from Camberley would probably not have blown her holiday money on a turban box. Dr Wakefield, a British visitor to Kashmir in 1875, took a particular interest in papier mâché work, which he described as 'both curious and elegant'; its main customers, he wrote, were 'the princes, the nobility and the wealthy ones of the land' who not only employed painters to decorate their houses, but ordered bedsteads, tables and chairs, as well as turban boxes, candlesticks, bracelet boxes, vases, bows and arrows, musical instruments, book covers, book stands (for the Koran), and screens. Maharaja Ranbir Singh, the ruler of Kashmir at the time of Dr Wakefield's visit, was an enthusiastic patron of the art who used to order coffee sets to present to his European friends. Large shallow papier mâché boxes were made in the nineteenth century especially for storing Kashmir shawls. These became very popular in France where, it seems, they fetched a high price even when sold separately from the shawls.

As the number of European visitors to Kashmir grew, the range of articles commissioned from the papier mâché dealers extended to include things dear to sahibs' and memsahibs' hearts—cigar and cigarette boxes, card cases, fan and glove boxes, knitting needle cases, picture frames, bridge-scoring pads, mirrors, desk sets, lamps of all sorts, overmantels, cotton wool boxes, powder bowls, calendar backs, and so on. The changing fashions in Europe were sometimes reflected in the papier mâché painting. For example, in the thirties when chrome became all the rage, as well as that particular thirties' shade of green that you sometimes see in old bathrooms—silver and green became favourite colours in Kashmir too. Suffering Moses says that the lovely pattern of all-blue flowers on a white background was inspired by Queen Mary's famous love of the colour blue. But Lockwood Kipling thought British taste very destructive: 'In response to the English demand for "something chaste" the rich colours and bold patterns formerly in vogue have given way to a somewhat sickly monochrome of cream colour and gold', he wrote. A dreary cream and gold painted bedroom suite of wardrobe, bedhead and dressing table that I once saw in an antique shop in London exactly fitted this description. It would have looked at home with a boring candlewick bedspread but never with a Kashmiri shawl.

Apart from what an object is—which can be a clue as to when it was made but is not always—the only other indication of the age of a papier mâché piece is the quality of the painting on it: the finer the brush strokes, the more detailed the pattern, the subtler the shading, the gentler the tones and the more gold leaf used, particularly on the background, the older it is likely to be. It is sadly revealing to compare present-day papier mâché work with the nineteenth century pieces in the Victoria and Albert Museum, the Lahore Museum, or the little Srinagar State Museum. (This has been housed since the early 1900s in an attractive, if shabby, building called the Lal Mandi, which was originally built in the last century to be a royal guest-house and banqueting hall.)

Even by the end of the last century admirers of the art of papier mâché painting were worried that standards were falling. In 1895 Sir Walter Lawrence, the British civil servant who helped sort out Kashmir's taxation system, wrote:

> Papier mâché has perhaps suffered more than any other industry from the taste of foreigners.... Ask an old artist in papier mâché to show the work which formerly went to Kabul and he will show something very different from the miserable trash which is now sold. But the Pathans of Kabul paid the price of good work; the visitors to the valley want cheap work and they get it.

Sir George Watt, writing in the catalogue for the Indian Art exhibition at Delhi in 1903, felt much the same. He was particularly concerned that so many so-called papier mâché objects were, in fact, made of wood, and mourned the days when papier mâché was composed of layers of paper separately applied to a mould; old coffee cups made in this way were so strong that the hottest brew could be drunk from them. Sir George was surprised to find while in Kashmir that

> neither in the Palace nor in the State Museum have there been preserved samples of the fine old forms of Kashmir papier mâché so much appreciated by collectors of Indian art. So completely have these disappeared that it would not be far from correct to affirm that when Kashmir thinks of reviving its former beautiful art of papier mâché, it will have to go to the museums of Europe and America for the most desirable models.

If Sir Walter and Sir George had fears for the industry then, how distraught they would have felt in the years since. For despite their warnings, the situation did not improve: it became worse. The traditional local customers—the princes and nobles of the land—fell away because it became more fashionable to choose European furnishing ideas than local ones. For instance, when Maharaja Hari Singh built himself the Gulab Bhawan palace (now the Oberoi Hotel) in 1932, he had it decorated with plaster mouldings rather than papier mâché. In the mean time the western visitors who were prepared to set money aside for the purchase of 'works of art', as Mrs Burrows had suggested in her 1895 guidebook, were soon hugely outnumbered by those simply wanting cheap souvenirs. All types of Kashmiri craftsmen suffered, not just the papier mâché makers. With rare exceptions their work degenerated, and their reputation declined, until many people came to think of those Kashmiri crafts, which for centuries had been admired and sought after by connoisseurs all over the world, as mere tourist rubbish.

And then came a new blow. Buyers from large stores in India, Europe and America began to shop in Kashmir. Like the tourists they wanted quick, cheap work, but on an enormous commercial, damaging, scale. While we were in Kashmir, a dealer friend received an order for 30,000 pill boxes to cost not more than Rs 3 each. This is a ridiculously low amount which should barely cover the cost of a hand-made papier mâché box, let alone any painting done on it. No wonder in some factories they work with fingertips dipped in colour. There is not enough time or money allowed to do it any other way.

The work of Kashmiri craftsmen was so poorly regarded and unfashionable that until quite recently carved wooden screens, fine enamel work, chain-stitched rugs and carpets, glowing papier mâché tables and boxes, fabulous shawls and embroideries, could be found in the junk shops of Europe selling for a pittance, and a person would think nothing of hacking up a Kashmir shawl to make cushions or clothing out of it. But now it is nearly half a century since the British left India and distance has lent enchantment. Generations have grown up to whom the story of the British raj is an enthralling, romantic, costume drama played against wonderfully exotic scenery. And suddenly the stage props—the Kashmiri screens, copper trays, painted lamps, inlaid tables, carved teak chairs from Burma and so on—that once furnished so many British Indian bungalows, are attracting the attention of collectors and dealers and discerning private buyers—old Kashmiri work of any sort is likely to be expensive nowadays.

One can only hope that this will be reflected in a new appreciation of today's craftsmen in Kashmir, that customers, aware that fine work can be an investment, will encourage the remaining artists in their best work. For unless we prize quality and excellence, and are prepared to pay a fair price for them, it is probable that one day all the old Kashmiri skills will, like shawl weaving, be forgotten. 'It is haggling and hurry that have spoiled art in Europe, and are spoiling it in Asia', wrote Sir George Birdwood in 1880. What an irony it would be if Kashmir's crafts had survived their terrible tortured history only to be destroyed by the meanness and mediocre taste of our own affluent society.

Principal Sites in Srinagar and the Kashmir Valley

(See the maps and keys to find locations)

Mughal Gardens

Shalimar and *Nishat* are the two best-known and the most classically laid out gardens. Shalimar has Shah Jahan's lovely black marble pavilion, Nishat's terraces are steeper. *Chashma Shahi* is a small, pretty garden halfway up a hill with marvellous views over the Dal lake, and gushing spring water that is famous for its purity. *Pari Mahal* is the most romantic of all the gardens near Srinagar. Built high on the shoulder of a hill, it has spectacular views, and the added interest of some reasonably intact buildings, including a dove cote. Its terraces are positively cliff-like, supported by the original arched walls in brick and stone.

Some distance from Srinagar, *Achabal* is a particularly watery garden with a powerful waterfall and canals and pools; there are lots of trees and it would be a pleasant place to take a picnic. Still further from the capital is *Verinag*, the garden that Emperor Jahangir loved best of all. All that is left of the palace he built there is a deep octagonal pool that contains the spring, around which goes an arcaded walkway in stone. Empress Nur Jahan had gold rings put through the noses of the fish in this pool.

It has to be said that all the Mughal gardens would greatly benefit from being replanted and restored with more flair, imagination and sensitivity.

There were other gardens at *Jarogha Bagh* on the Manasbal lake, at *Dilawar Khan Bagh* in Srinagar (where some of the eighteenth and nineteenth century European travellers camped), at *Bijbihar* on the way to Anantnag, on the Sona Lank or *Char Chinar* island in the Dal lake, at *Nasim Bagh* near Hazratbal, and in many other places. But of these gardens there is barely a trace left today—only a few old chenar trees are left standing. At *Kokernag* there is a pleasant new garden.

Hindu Temples

Shankaracharya temple dominates Srinagar. Its foundations are said to go back to Emperor Ashoka's time, but it has been ruined and rebuilt several times in the centuries since. *Martand* and *Avantipur*, now in ruins, were the grandest of the Hindu temples built between the second and tenth centuries AD, and there are other temples or remains at *Patan*, *Wangat*, *Payer*, *Bhaniyar*, and at *Mamalsvara* near Pahalgam. There is a partially submerged temple in the *Manasbal lake*, and a curious cave temple in the cliffs just outside *Mattan*, to the right of the road as you drive to Pahalgam. The best-preserved temple and the easiest to get to is the one at *Pandrethan*, on the road out of Srinagar towards Anantnag

Mosques

The *Akhund Mullah Shah mosque* was built by Prince Dara Shikoh for his spiritual tutor. It is nicely carved in grey limestone, but sadly neglected: well worth seeking out if you are interested in the Mughals and their architecture. So is the elegantly proportioned *Pathar Masjid* built by Empress Nur Jahan in Srinagar city. The *Jama Masjid* in Srinagar is famous for its forest of huge, straight pillars made of the trunks of Deodar trees. *Shah Hamadan* mosque is built out of wood to look as though it were brick. It is intricately painted inside, and has a pagoda-like roof and metal spire. The mosque at *Hazratbal* is new (built on the site of an old one). It contains a hair of the Prophet.

Ruins

There are old sites and ruins all over Kashmir. Ancient paving has been unearthed near Harwan, there are relics of a city (built by King Lalitaditya) near Patan, traces of old water works near Shadipur, and many other remains of temples and shrines. A great deal of recycling of stone has gone on for centuries—huge blocks from temples being used, for instance, in the embankments of the Jhelum in Srinagar. At the foot of Hari Parbat hill (on the city side) was the city of *Nagar Nagar* which Akbar built—two of its great gates are still standing and reasonably intact. On Lanka island in the Wular lake are the ruins of Sultan Zain-ul-Abidin's shrine.

Srinagar

The roofs of the old wooden houses of Srinagar used to be covered with earth and plants and flowers—they are now most often covered with corrugated iron, but Srinagar is still an extraordinarily medieval-looking city and a most interesting one to stroll around. It is a terrible pity, though, that some effort has not been made to construct new buildings in the traditional Kashmiri style of bricks and wood. The new hotels that have sprung up along the Boulevard on the Dal lake are concrete blocks, and the new government buildings could be in any dreary town anywhere in the world. A fairly recent tragedy was the filling-in of the very old Mar Nullah canal to make a road. This narrow, ancient canal could be a smelly place, but the houses on either side of it, and the stone embankments, were some of the oldest and quaintest in the city.

The bridges of Srinagar were made of deodar wood, and some had wooden shops along them. The shops have gone now, but a few of the old bridges with their massive scaffolding-like wooden supports remain. Some new bridges have been built, so there are now ten in all. Starting with Zero Bridge (a new one), they go downstream in this order: Amira Kadal, Badshah Kadal, Haba Kadal, Fateh Kadal, new Fateh Kadal (also called Biscoe Bridge because it is near the site of Tyndale-Biscoe's mission school), Zaina Kadal, Ali Kadal, Nava Kadal, Safa Kadal.

Walks

In and around Srinagar there are all sorts of walks to do—from strolling along the Bund, to crossing the Dal lake on the Suttoo (causeway), to climbing up to Pari Mahal (down would be less arduous) or Chashma Shahi, or to the tops of Shankaracharya or Hari Parbat hills. Not far from Srinagar are green foothills, often rich in wild flowers, and there is a game reserve at Harwan.

Bibliography

CHAPTER ONE: HISTORY AND GENERAL INFORMATION

Rajatarangini by Kalhana, translated by M. A. Stein, London, 1900.

Kashmir by G. M. D. Sufi. University of Punjab, 1949.

Akbar-nama by Abul-Fazal, translated by H. Beveridge. Asiatic Society. Calcutta, 1939.

Ain-i-Akbari, translated by H. Blochmann and H. S. Jarrett, Calcutta, 1873–94.

Tuzuk-i-Jahangiri, translated by A. Rogers. Royal Asiatic Society. London, 1909–14.

Remonstrantie by Francis Pelsaert, translated by W. H. Moreland (as *Jahangir's India*), Cambridge, 1925.

Cashmere Misgovernment by Robert Thorp, Calcutta, 1868.

Illustrations of Ancient Buildings in Kashmir by H. H. Cole, 1869.

The Northern Barrier of India by Frederic Drew, 1877.

The Happy Valley by W. Wakefield, M.D., 1879.

Hyderabad, Kashmir, Sikkim and Nepal by Sir R. Temple, 1887.

Where Three Empires Meet by E. F. Knight, London, 1893.

The Vale of Kashmir by W. Lawrence, 1896.

Memoirs of A. Gardner, Edinburgh, 1898.

Kashmir by Sir Francis Younghusband. A. and C. Black, 1917.

Kashmir Handbook by Dr Duke. Revised edition, 1903. Thacker, Spink and Co., Calcutta.

The Tourists Guide to Kashmir, Ladakh, Skardo, etc. by Major Arthur Neve. Revised edition by Dr E. F. Neve, 1923.

Gardens of the Great Mughals by Constance Villiers Stuart, A. and C. Black, 1913.

The Gardens of Mughul India by S. Crowe and S. Haywood. London, 1972.

The Mr A. Case, edited by C. E. Bechhofer Roberts. Jarrolds, London, 1950.

The Great Moghuls by Bamber Gascoigne, Jonathan Cape, 1971.

CHAPTER TWO: TRAVELLERS

Yuan-Chwang's Travels in India (629–645 AD) by Thomas Watters, Royal Asiatic Society, London, 1904.

Akbar and the Jesuits by Father Pierre du Jarric, translated by C. H. Payne, 1926.

Jahangir and the Jesuits by Father Fernao Guerreiro, translated by C. H. Payne, 1930.

Journey from Rome to Lhasa by Father Desideri. Lettres edifiantes, 1716.

The Embassy of Sir Thomas Roe to India, 1615–19 edited by W. Foster, London, 1926.

Travels in the Moghul Empire by Francois Bernier, translation by Irving Brock, 1826.

Journey from Bengal to England by G. Forster, 1798.

Letters from India; describing A Journey in the British Dominions of India, Tibet, Lahore, and Cashmere, during the years 1828, 1829, 1830, 1831, by Victor Jacquemont, London, 1834.

Travels in the Himalayan provinces of Hindustan and the Panjab; in Ladakh and Kashmir, in Peshawar, Kabul, and Munduz Bokhara, 1819–1825 by William Moorcroft, edited by G. Trebeck. London, 1841.

Travels in Kashmir, Ladak, Iskardo, etc. by G. T. Vigne, London, 1842.

Travels in India and Kashmir by Baron von Schonberg, 1843–4.

Travels in Kashmir and the Panjab by Baron C. Hügel, London, 1845.

Kashmir Panorama. Description of a view of the valley and city of Cashmere now exhibiting at The Panorama, Leicester Square. From drawings by G. T. Vigne, 1849.

Researches and Missionary Labours by the Rev. Joseph Wolff, London 1835.

Lalla Rookh by Thomas Moore, 1861.

CHAPTER THREE: THE BRITISH IN KASHMIR

The Adventures of a Lady in Tartary, Thibet, China and Kashmir by Mrs Hervey, 1853.

Novels by Mrs Hervey: Snooded Jessaline and Margaret Russell.

The Journals of Honoria Lawrence. India Observed, 1837–1854. Edited by John Lawrence and Audrey Woodiwiss. London, 1980.

Honoria Lawrence. A Fragment of Indian History by Maud Diver, 1936.

Summer Scenes in Cashmere by Mrs H. Clark, 1858.

Diary of a Pedestrian in Cashmere and Thibet by Captain Knight, Forty-eighth Regiment, 1863.

Wanderings of a Naturalist in India, the Western Himalayas, Cashmere by Andrew Leith Adams, 1867.

Our Visit to Hindostan, Kashmir and Ladakh by Mrs J. C. Murray Aynsley, 1879.

Irene Petrie, Missionary to Kashmir by Mrs Ashley Carus-Wilson, 1900.

Cashmere En Famille by Mrs Burrows, 1902.

A foot through the Kashmir Valleys by Marion Doughty, 1902.

A Lonely Summer in Kashmir by Margaret Cotter Morrison, 1904.

The Emerald Set with Pearls by Florence Parbury, 1909.

Thirty Years in Kashmir by Arthur Neve, London, 1913.

Picturesque Kashmir by Arthur Neve and Geoffroy Millais, London, 1900.

Cashmere: Three weeks in a Houseboat by A. Petrocokino, London, 1920.

Tyndale-Biscoe of Kashmir. An autobiography, 1951.

CHAPTER FOUR: ARTS AND CRAFTS

Some of the general books listed under Chapter One contain information about the various arts and crafts of Kashmir, and many of the travellers whose books are listed under Chapter Two wrote descriptions of them, particularly Francois Bernier, George Forster, and—most detailed and valuable of all—William Moorcroft. Additional sources are:

Travels in India and Kashmir by Baron von Schonberg, 1843–4.

Kashmeer and its Shawls, 1875.

Lady Login's Recollections, 1820–1904 by E. Dalhousie Login, 1917.

Shawls by John Irwin, Her Majesy's Stationery Office, 1955.

The Kashmir Shawl by John Irwin, Her Majesty's Stationery Office, 1973.

Letters of Laurence Sterne edited by L. P. Curtis, Oxford, 1935.

Marg, a magazine of the arts. March 1955, volume VIII.

Kashmir to Paisley by Caroline Karpinski, from the Metropolitan Museum of Art Bulletin, November 1963.

The Indian Collections: 1798–1978 by Robert Skelton, Burlington Magazine, May 1978.

Lacquerwork in Asia and Beyond by Zebrowski and Watson. Percival David Colloquy papers, no. 11. The Foundation (London University), 1982.

Miniatures from Kashmirian Manuscripts by A. Adamova and T. Greck.

CATALOGUES

Great Exhibition of the Works of Industry of all Nations, 1851. Official descriptive and illustrated catalogue.

International Exhibition, 1862. Classified and descriptive catalogue by J. F. Watson.

Official catalogue of contributions from India to the Exhibition of 1862 by A. M. Dowlean.

Punjab Exhibition, 1864.

Catalogue of the Objects of Indian Art Exhibited in the South Kensington Museum by Henry Hardy Cole, 1874.

The Paris Universal Exhibition of 1878. Handbook to the Indian Court by Sir George Birdwood, revised as *The Industrial Arts of India, a South Kensington Museum Art Handbook*, 1884.

Colonial and Indian Exhibition, 1886.

Colonial and Indian Exhibition, 1886. Catalogue of exhibits by Govt of India.

Art Manufactures of India, specially compiled for the Glasgow International Exhibition, 1888 by T. N. Mukharji of the Indian Museum, Calcutta.

The Empire of India Exhibition, 1895. Official catalogue.

Indian Art at Delhi, 1903.

British Empire Exhibition, 1924. Report by the Commissioner for India for the Exhibition.

Examples of Indian Art in the British Empire Exhibition, 1924.

The Art of the Shawl. Catalogue for an exhibition of woven and printed shawls. West Surrey College of Art and Design, 1977.

The Islamic Perspective, World of Islam Festival Trust. Catalogue for exhibition at Leighton House, London, 1983.

Index

MORE OXFORD PAPERBACKS

HISTORY IN OXFORD PAPERBACKS

As the Oxford Paperbacks' history list grows, so does the range of periods it covers, from the Pharaohs to Anglo-Saxon England, and from Early Modern France to the Second World War.

EGYPT AFTER THE PHARAOHS

Alan K. Bowman

The thousand years between Alexander the Great's invasion in 332 BC and the Arab conquest in AD 642 was a period of enormous change and vitality in the history of Egypt. The Hellenistic era under the powerful Ptolemies ended with the defeat of Antony and Cleopatra in 30 BC, and Egypt became a province of Rome.

Throughout the millenium, however, many of the customs and belief of old Egypt survived, adapting themselves to the new rulers, who were in turn influenced by Egyptian culture. The heritage of the Egypt of the Pharaohs remained a vital force in the history of the land until the coming of Islam.

A vast collection of papyrus texts has survived from this period recording not only the great events but the everyday letters, lawsuits, accounts, and appeals of ordinary Egyptians. From these texts and from the evidence of archaeology, Dr Bowman draws together the Egyptian, Greek, and Roman strands of the story, presenting a masterly survey of the history, economy, and social life of Egypt in this thousand year span.

'eminently readable . . . should be studied by anyone who is seeking details of everyday life in the Roman period' *British Archaeological News*

Also in Oxford Paperbacks:

A History of the Vikings Gwyn Jones
A Turbulent, Seditious, and Factious People Christopher Hill
The Duel in European History V. G. Kiernan

OXFORD BOOKS

Oxford Books began in 1900 with Sir Arthur Quiller-Couch ('Q')'s *Oxford Book of English Verse.* Since then over 60 superb anthologies of poetry, prose, and songs have appeared in a series that has a very special place in British publishing.

THE OXFORD BOOK OF ENGLISH GHOST STORIES

Chosen by Michael Cox and R. A. Gilbert

This anthology includes some of the best and most frightening ghost stories ever written, including M. R. James's 'Oh Whistle, and I'll Come to You, My Lad', 'The Monkey's Paw' by W. W. Jacobs, and H. G. Wells's 'The Red Room'. The important contribution of women writers to the genre is represented by stories such as Amelia Edwards's 'The Phantom Coach', Edith Wharton's 'Mr Jones', and Elizabeth Bowen's 'Hand in Glove'.

As the editors stress in their informative introduction, a good ghost story, though it may raise many profound questions about life and death, entertains as much as it unsettles us, and the best writers are careful to satisfy what Virginia Woolf called 'the strange human craving for the pleasure of feeling afraid'. This anthology, the first to present the full range of classic English ghost fiction, similarly combines a serious literary purpose with the plain intention of arousing pleasing fear at the doings of the dead.

'an excellent cross-section of familiar and unfamiliar stories and guaranteed to delight' *New Statesman*

Also in Oxford Paperbacks:

The Oxford Book of Short Stories edited by V. S. Pritchett
The Oxford Book of Political Anecdotes
edited by Paul Johnson
The Oxford Book of Ages
edited by Anthony and Sally Sampson
The Oxford Book of Dreams edited by Stephen Brock

OXFORD LIVES

Based on original research and written in each case by acknowledged experts on their subject, the authoritative accounts included in the Oxford Lives series are the best biographies available.

RODIN

Frederic Grunfeld

Until the publication of this book, Auguste Rodin, the greatest sculptor of the nineteenth century, had been notoriously ill-served by his biographers. 'All the stuff written about him' commented George Bernard Shaw 'is ludicrous cackle and piffle.'

Frederic Grunfeld's original and exhaustive research has enabled him to penetrate the legends perpetrated by Rodin's earlier hagiographers and detractors. For the first time a clear picture of Rodin's life emerges: his impoverished youth, his many love-affairs, his friendships and enmities, his public life, and above all his devotion to art and the life-long struggle to achieve official recognition without compromise.

The age in which Rodin lived is also vividly conveyed. After his early successes Rodin was surrounded by the wealthy and famous. Zola, Balzac, Rilke, R. L. Stevenson, Edith Wharton, Debussy, Oscar Wilde, Edward VII, Clemenceau, and Kaiser Wilhelm II are just some of Rodin's famous contemporaries who figure in his life story, and without whom the story would not be complete.

'Frederic Grunfeld has written a biography which by any normal standards must be called definitive. It is deliciously crammed with information . . . Rodin's erotic obsessions are explored with a degree of tact rare among modern block-buster biographies.' *Sunday Telegraph*

Also in Oxford Lives:

Clara Schumann: The Artist and the Woman Nancy B. Reich
Gladstone 1809–1874 A. C. G. Matthew
Nikolai Gogol Vladimir Nabokov
James Joyce Richard Ellmann

OXFORD REFERENCE

Oxford is famous for its superb range of dictionaries and reference books. The Oxford Reference series offers the most up-to-date and comprehensive paperbacks at the most competitive prices, across a broad spectrum of subjects.

THE CONCISE OXFORD COMPANION TO ENGLISH LITERATURE

Edited by Margaret Drabble and Jenny Stringer

Based on the immensely popular fifth edition of the *Oxford Companion to English Literature* this is an indispensable, compact guide to the central matter of English literature.

There are more than 5,000 entries on the lives and works of authors, poets, playwrights, essayists, philosophers, and historians; plot summaries of novels and plays; literary movements; fictional characters; legends; theatres; periodicals; and much more.

The book's sharpened focus on the English literature of the British Isles makes it especially convenient to use, but there is still generous coverage of the literature of other countries and of other disciplines which have influenced or been influenced by English literature.

From reviews of *The Oxford Companion to English Literature Fifth Edition:*

'a book which one turns to with constant pleasure . . . a book with much style and little prejudice' Iain Gilchrist, *TLS*

'it is quite difficult to imagine, in this genre, a more useful publication' Frank Kermode, *London Review of Books*

'incarnates a living sense of tradition . . . sensitive not to fashion merely but to the spirit of the age' Christopher Ricks, *Sunday Times*

Also available in Oxford Reference:

The Concise Oxford Dictionary of Art and Artists
edited by Ian Chilvers
A Concise Oxford Dictionary of Mathematics
Christopher Clapham
The Oxford Spelling Dictionary compiled by R. E. Allen
A Concise Dictionary of Law edited by Elizabeth A. Martin

OXFORD LETTERS & MEMOIRS

This popular series offers fascinating personal records of the lives of famous men and women from all walks of life.

JOURNEY CONTINUED

Alan Paton

'an extraordinary last testament, told in simple and pungent style . . . for anyone new to the period and to Paton, it will be a revelation' *Independent*

This concluding volume of autobiography (the sequel to *Towards the Mountain*) begins in 1948, the year in which Paton's bestselling novel, *Cry, the Beloved Country*, was published, and the Nationalist Party of South Africa came to power. Both events were to have a profound effect on Paton's life, and they represent two major themes in this book, literature and politics.

With characteristic resonance and trenchancy, Paton describes his career as a writer of books, which were received with extreme hostility by his fellow South Africans, and also covers his political life, notably the founding—and later Chairmanship—of the Liberal Party of South Africa, the multi-racial centre party opposed to apartheid.

'required reading for anyone who wants to understand, compassionately, the full tragedy of South Africa' *Daily Express*

Also in Oxford Letters & Memoirs:

Memories and Adventures Arthur Conan Doyle
Echoes of the Great War Andrew Clark
A Local Habitation: Life and Times 1918–1940
Richard Hoggart
Pack My Bag Henry Green

ILLUSTRATED HISTORIES IN OXFORD PAPERBACKS

Lavishly illustrated with over 200 full colour and black and white photographs, and written by leading academics, Oxford Paperbacks' illuminating histories provide superb introductions to a wide range of political, cultural, and social topics.

THE OXFORD ILLUSTRATED HISTORY OF ENGLISH LITERATURE

Edited by Pat Rogers

Britain possesses a literary heritage which is almost unrivalled in the Western world. In this volume, the richness, diversity, and continuity of that tradition are explored by a group of Britain's foremost literary scholars.

Chapter by chapter the authors trace the history of English literature, from its first stirrings in Anglo-Saxon poetry to the present day. At its heart towers the figure of Shakespeare, who is accorded a special chapter to himself. Other major figures such as Chaucer, Milton, Donne, Wordsworth, Dickens, Eliot, and Auden are treated in depth, and the story is brought up to date with discussion of living authors such as Seamus Heaney and Edward Bond.

'[a] lovely volume . . . put in your thumb and pull out plums' Michael Foot

'scholarly and enthusiastic people have written inspiring essays that induce an eagerness in their readers to return to the writers they admire' *Economist*

Other illustrated histories in Oxford Paperbacks:

The Oxford Illustrated History of Britain
The Oxford Illustrated History of Medieval Europe